OF COCKROACHES
AND CRICKETS

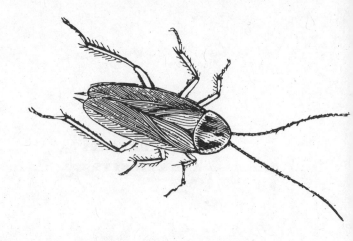

FRANK NISCHK

Foreword by **CARL SAFINA**

Translated by **JANE BILLINGHURST**

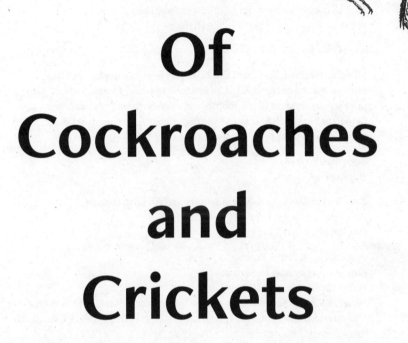

Of Cockroaches and Crickets

Learning to Love Creatures That Skitter and Jump

GREYSTONE BOOKS
Vancouver/Berkeley/London

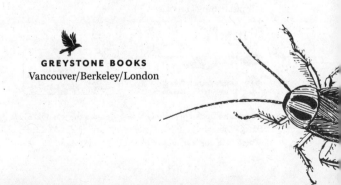

First published in English in 2023 by Greystone Books
Originally published in German as *Die fabelhafte Welt der fiesen Tiere: Von fürsorglichen Schaben, tauchenden Libellen und boxenden Krebsen— Eine Liebe auf den zweiten Blick* by Frank Nischk © 2020 Ludwig Verlag, part of the Random House GmbH publishing group, Munich
English translation copyright © 2023 by Jane Billinghurst
Foreword copyright © 2023 by Carl Safina

23 24 25 26 27   5 4 3 2 1

Greystone Books Ltd.
greystonebooks.com

Cataloguing data available from Library and Archives Canada
ISBN: (cloth) 978-1-77164-872-1
ISBN: (epub) 978-1-77164-873-8

Copyediting by James Penco
Proofreading by Meg Yamamoto
Jacket and text design by Fiona Siu
Jacket images by Natchanon Kongsakul (cockroach), GraphicsR F.com (cricket) / Shutterstock.com; antagain (grasshopper), Dewin ' Indew (leaves), draganab (background) / iStock.com
Interior images by Morphart Creation (cockroach), Hein Nouwens (cricket), antlexx (butterfly) / Shutterstock.com

Printed and bound in Canada on FSC® certified paper at Friesens. The FSC® label means that materials used for the product have been responsibly sourced.

Greystone Books thanks the Canada Council for the Arts, the British Columbia Arts Council, the Province of British Columbia through the Book Publishing Tax Credit, and the Government of Canada for supporting our publishing activities.

Canada

Greystone Books gratefully acknowledges the xʷməθkʷəy̓əm (Musqueam), Sḵwx̱wú7mesh (Squamish), and səlilwətaɬ (Tsleil-Waututh) peoples on whose land our Vancouver head office is located.

For Martin Dambach

Contents

PART III: GIVING NATURE A FIGHTING CHANCE

Foreword

"They're sooo *disgusting*."

"They carry disease."

"They kill trees."

"They can hurt you."

"Who needs 'em?"

No, we're not talking about people. We're talking about insects. And it's true. It often seems they are all bad. Except...

Except, let's answer that last question about "who needs them." Flowering plants, for starters. Human—among other— beings. And an awful lot in between, such as, well, birds, amphibians, freshwater fish, and on and on. Without insects who eat insects, there'd be a lot more insects. So, you see, it gets complex, and sweeping generalizations don't help—except to say, insects make the world go round. We can live with them, but we couldn't live without them.

And by the way, have you ever *seen* insects? I mean, have you ever really *looked*? There are universes on six legs all around us. A few engage in some *very* nasty cloak-and-dagger strategies that endanger life, but 98 percent (or maybe more) live glorious lives that engender life as well. This book *is* a look. A deep look that will change your view by giving you a bigger picture of insects than you've ever had. Or even imagined.

~~~~~~

AND SPEAKING OF *Cockroaches and Crickets*, I'm *not* one to talk. When I was five years old and lived in the city—in Brooklyn, New York—a miraculous thing happened one summer. A cricket! One night I heard a cricket calling. What a beautiful sound. It was the only one, a lonely soloist calling for a mate. Each evening I'd lie in bed listening to the cricket. A few nights after the serenading began, I happened to be outside with my father shortly after dark. I saw a big cockroach and dutifully stepped on it without a thought, as I'd been taught to do.

"Dad, look at this big roach I just killed!" So proud.

"That's not a roach, Carl; that's a cricket." I had killed the night's music. Whoa.

After that, I learned to crouch before I crushed, to at the very least have a second look. And just the other day as my wife and I were doing a deep clean of our house, I saw her go to the door several times to let spiders go outside. It feels good to pay attention. And, at least most of the time, to live and let live. I would not admit this to just anyone, or to my wife (though she knows), but I vacuum *around* most spiders. I figure if I don't want a lot of bugs in the house, well, it's okay to have a few spiders there to eat them. It's a deal.

Crouching for a look is a good first start. But this book does a lot more. Have you ever seen the delicate veins in a cockroach's wing? Have you stared lovingly into their art-like array of complex spherical compound eyes? This book will take you deep into that mutual gaze, seeing things no human eye can see without the aid of tools such as scanning electron microscopes. And would you ever have guessed that some cockroaches not only care for their young but feed them their own kind of *milk*?

There's a lot to think about. The smaller the worlds we can see, the bigger our world becomes. This is a horizon-expanding book, that's for sure.

And now, to make it easy, all you have to do is turn the pages, savor the exquisite writing, and world upon world will be revealed. You'll be a bug lover in no time.

—CARL SAFINA

Beyond the Cute and Cuddly

"The lord of rats and eke of mice,
Of flies and bedbugs, frogs and lice..."
JOHANN WOLFGANG VON GOETHE, *Faust*, scene 3

"Frankie, come quickly! There's a monitor lizard outside!" My mother's voice rang through the house, immediately putting me on high alert. The image of an enormous carnivorous reptile slinking through my father's tulips leapt into my adolescent brain. I ignored the improbability of this thought and rushed to the window.

"Where? Where? WHERE IS IT?"

I scanned every corner of the yard. My researcher's heart was beating wildly.

Reptiles were my thing when I was young. I was never that interested in ponies and I never longed for a dog as a playmate. I was immune to the charms of bright-eyed puppies, wobbly kittens, or fluffy hamsters. As far back as I can remember, my favorite place to hang out was the huge terrarium in the aquarium at the Cologne Zoo where the monitor lizards lived. I could have spent hours watching the lanky giants spread languorously on the concrete floor of their German home. Every once in a while, one of them took a dip in the enclosure's tank—that was about it for action. And yet I could not tear myself away. My parents, and my brother, who was two years older than me, thought I was an annoying little weirdo, and thanks to my obsession, family visits to the zoo became something to be endured rather than enjoyed.

We lived in Wesseling at the time, a little town on the outskirts of Cologne. "Town" was then, and is now, a generous description of this settlement on the Rhine. Huge chemical plants and oil refineries surrounded the approximately thirty thousand inhabitants. Silos, chimneys, and flare stacks towered over the tallest buildings. At night, thousands of neon lights shining from the chemical plants and flares burning from the refineries made the whole thing look a little like the skyline of Manhattan. By day, unfortunately, it reverted to what it was: an ugly industrial area.

~~~~~~

WE LIVED IN a small estate of workers' houses that ended at the gates of one of the chemical plants. Both the plant and the estate were a stone's throw from the Rhine. All the children were strictly forbidden to go anywhere near the river, which in those days was basically an open sewer. Although the water quality in the Rhine has been steadily improving thanks to ever-more-efficient sewage treatment facilities, there was one thing that was better back then than it is today. Our small industrial town, tucked in next to enormous factories, was full of front and backyards left to run riot, where a young boy mad about nature could discover something new every day.

House martins raised their chicks in mud nests stuck to the walls under the eaves of our modest home, but these summer guests were so ordinary that I barely gave them a second glance. My mini-voyages of discovery led me to more exciting things.

Chirping grasshoppers, for instance. When I ran through the wild spaces with my friends, grasshoppers sprang out from lush flowery meadows. We caught them and, I am sorry to say, subjected them to gruesome experiments. What would happen to a grasshopper, we wondered, if we threw it into one of the many spiders' webs? Web-spinning spiders were almost as numerous in the workers' estate as grasshoppers, beetles, bugs, and other insects.

But an exotic lizard? In our yard? My desire to see my favorite animal in the wild was so great that I *wanted* to hear the words "monitor lizard" when my mother called me over. In German, the word for monitor lizard is *Waran* and the word for pheasant is *Fasan*; my mother is Austrian and when she speaks, a hard *F* sometimes sounds like a *W*, and vice versa. As I was pestering her excitedly, wanting to know where the beast was, she pointed to a large bird with an impressively long tail strutting over the grass. This was no monitor lizard; it was a pheasant, more precisely a golden pheasant, that had wandered into the estate. I was hugely disappointed. To this day, beautiful though they are, pheasants just don't do much for me.

~~~~~~~

FIFTY YEARS LATER, as I walk through the streets I roamed as a child, I see how everything has changed. The old workers' houses still stand, but the once-overgrown yards have been either built on or asphalted over to serve as parking spaces. Robot lawn mowers patrol the grass. There's little chance a pheasant would stray into this section of Wesseling anymore. Most of the insects have vanished as well—as have the spiders and house martins, now that their juicy prey is gone.

So what? you might ask. What's the big deal about a few grasshoppers?

My childhood memories highlight a couple of issues. Slowly but surely, it's beginning to dawn on more than just scientists that we are in the midst of an era of dramatic species decline. Tigers, rhinoceroses, and gorillas are not the only animals in danger. Tens of thousands of other species, most barely larger than your thumbnail, are quietly disappearing. We notice only when we remember the small hunting expeditions of our childhoods.

The other thing you need to know is that I am a nerd, a devotee of insects. That means I'm interested in small creatures unloved by many. It's easy to love ponies and pandas. Baby faces, button eyes, and soft fur—this is usually why we fall in love with animals. Biologists who specialize in appealing animals often focus on their humanlike characteristics, the idea of the best friend—perhaps even the better person. But what about leeches, cockroaches, sea cucumbers, and nematodes? And, of course, monitor lizards. Ninety-nine percent of living creatures are not standout stars, but each and every one of them can be just as interesting as a gigantic blue whale or an iridescent hummingbird. Even as a child, I did not want to be a biologist who focused on what was cute. I wanted to be an animal researcher who helped tell at least a few of the countless fascinating stories written by evolution—many of them works in progress to this day. That is why I became a biologist, why I spent a year studying the behavior of baby cockroaches, and why I am constantly returning to explore the rainforests of South America, where the incredible nightly chorus of a multitude of insects, frogs, and birds enthralls me anew every time I visit.

In this book, I want to tell stories that show how life finds a way. Inconspicuous critters that might bite, that we think of as disgusting and annoying if we think of them at all, are often the ones whose stories surprise us most. We look after those things we recognize, understand, and—perhaps—even love. And why shouldn't baby cockroaches be among those things?

PART I

The Year of the Cockroach

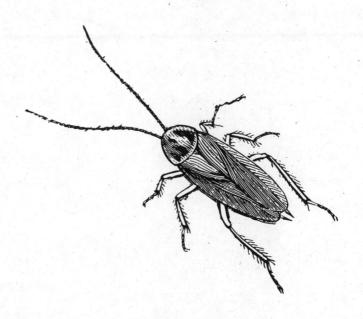

1

~

Cockroaches Instead of Hummingbirds

very journey has to begin somewhere. My journey
of exploration into the unloved animals of this planet
began in 1993 in a modest hotel room in Cali, Colombia,
a city with a population of over one million. I was nearing the
end of my undergraduate studies in biology at the University
of Cologne and I desperately needed a subject for my thesis. I
wanted to be a biologist, certainly, but I also wanted to explore
the world. Tropical rainforests had fascinated me since my
childhood, and I longed to visit the Amazon, the mightiest
river on Earth. Time and time again, I had traced its length
with my finger in my school atlas. And so I asked the biology
professors in Cologne: "Do any of you have colleagues in South
America who might be looking for a student?" I knew nothing
about tropical ecology, but I was super-enthusiastic.

I was fortunate enough to be put in contact right away with a
German-Colombian research team that wanted me as an intern.
An intern—well, that was something, at least. I learned that the

team had established a small research station in Colombia in one of the last undisturbed rainforests on the western side of the Andes. The station was located two hours by bus from Cali, which was the third-largest city in the country. The Amazon, of course, flows on the other, eastern side of the Andes, but they had me the moment I heard the word "rainforest."

When I heard about the biologists' research subject, I thought I had hit the jackpot. The research was all about hummingbirds, those tiny shimmering birds that flit from flower to flower whirring like miniature helicopters. Breathtaking aerial performers that can even fly backwards, they hover with no need to touch down as they sip nectar, their sugary fuel, from flowers. Back then I wasn't a birder. Indeed, I was only moderately interested in the world of birds, but I was more than willing to make an exception for the incandescent acrobats that were my ticket to the jungle.

~~~~~~

I ACCEPTED THE position immediately and bought my airline ticket to Colombia. I was excited about what the next few months would bring—with the exception of a few not-insignificant, and slowly rising, fears.

Back in the 1990s, there were few countries with a worse reputation than Colombia. Newspaper reports about this Andean country focused exclusively on violence: drugs, death squads, and guerrilla fighters. Everyone knew the name of the drug lord Pablo Escobar, but what was the name of the country's president? Not a clue. All the headlines about murder victims, kidnappings, and violence were indeed true, but what was then inconceivable for an average Central European like me was that Colombia also had universities, biologists, and citizens who led completely normal lives every day.

And so there I was, sitting in my cheap hotel room, not daring to venture out into the alien metropolis of Cali. But what kind of a tropical researcher was I if I couldn't even negotiate a concrete jungle? I devised a strategy to confront the problem head-on, and I fervently hoped it would help me overcome my fear. I was going to walk directly to Cali's busiest street. There, on Avenida Sexta, I would go to a restaurant and select a table next to the window, so I could observe day-to-day life on the street—from an appropriately safe distance—while I ate my lunch. I hoped that when I saw office workers, schoolchildren, and other normal people walking along the sidewalk, I would feel that I, too, could partake of normal daily life in Colombia. That, at least, was my hope.

At first, everything happened as planned. I was soon sitting in a nice little restaurant in front of a plate of rice, beans, and plantains, with a large grilled fish on top. As I had expected, ordinary people were out and about on the street. I began to relax, and I turned my attention to my lunch. There was one aspect of this experiment, however, that I had not thought through. Cali is a tropical city and because of the heat, the windows of the restaurant had no glass in them to protect me from the metropolis. It wasn't long before a homeless person noticed me. He shuffled over to the window, stretched his hand through the glassless window frame, and looked at me expectantly.

Was one supposed to give beggars something through a restaurant window in Cali? The man realized I was considering the matter. But I was taking too long, so he simply grabbed my fish and began eating it as he ambled away.

My plan to slowly acclimatize to life in Colombia had been derailed, at least for now. I will never survive this country, I thought.

LUCKILY, I WAS wrong. Just a few days later, I had fallen in love with the country, with Cali as a city, and with the amazing natural world in the tropics. I took the bus to Bajo Anchicayá, a small hydroelectric project on the old mountain road that led from Cali to the Pacific port of Buenaventura. Unlike many parts of the Andes, the hills and mountains here were still covered with dense rainforest. Thick rain clouds from the Pacific Ocean drifted in every afternoon. Western Colombia is one of the rainiest places in the world. Steep mountains and heavy rainfall are an ideal combination if you want to harness water to generate power. The hydroelectric station comprised a reservoir, a wooden hut to house staff and workers, workshops, a cafeteria, and the powerhouse, where turbines and generators hummed day and night. This was where the research team was staying as well: a biologist from Colombia, a graduate student from Germany, and three German interns. Rainforest stretched all around us as far as the eye could see.

~~~~~~

IN THE FOREST, they zipped from flower to flower: hummingbirds with names as scintillating as their plumage. Violet-tailed sylphs, western emeralds, and Andean emeralds buzzed around between the trees, nested, and raised their tiny chicks. I found myself living my childhood dream. I had never before been surrounded by so much beauty. The trees were hung with bromeliads and orchids. Tree ferns were as large as small palm trees. We saw long-billed toucans and listened to the calls of howler monkeys. I held my first boa constrictor, a snake almost one and a half times as long as I was tall that squeezes its prey to death. Everything seemed perfect.

But that was only half true. I had made it to the tropics, but I was a long way from being a tropical researcher. Instead of doing research, my fellow interns and I found ourselves caught between two sides of a quarrel. Biologists from the university in Cali and my German boss were accusing each other of failing to honor agreements. That was not good for us, because the Colombians seemed to have more leverage (and they also seemed to have the stronger case). The work permits for all Germans—whether professors, graduate students, or interns— were terminated overnight. And right then, my dream of finding a thesis topic in the rainforest died. We were allowed to continue living in paradise, but we were forbidden to work. We heard from Germany that the issues would likely be resolved and so we waited for better news. A week passed. Then two. Then three. At some point it dawned on me that I was wasting my time in the forests of Bajo Anchicayá. Although it was beautiful and the workers at the small hydroelectric station had treated us well, my plan had failed.

My work with hummingbirds is over, I thought. I still had two months until my flight home. Two months to be a tourist in Colombia: I swam in the Pacific Ocean, climbed volcanoes, and danced the salsa in Cali's *salsotecas*. Then I had to go back to Germany to plan for my future. It was clear that any aspect of observing hummingbirds in Colombia was completely out of the question as a thesis topic. It was time for plan B. My future rested on a sympathetic hearing from teams back at my university.

Martin Dambach was then one of the few professors at the University of Cologne still teaching classical ethology, the science of animal behavior. I wanted to study living animals and observe their interactions, so this sounded perfect. Yes, the gray-haired, whiskery-jawed zoologist told me, he had a thesis topic for me. And I was especially lucky because the work came with a paid position as a student assistant. To get paid for working on a thesis was extremely rare. It was tantamount to winning a lottery. Was there a catch? I asked.

To answer my question, the professor led me into a small room. There was a large red plastic container on the floor. To be more specific, the container was not on the floor, but floating in a tank full of water. The water, Dambach explained, was to make it more difficult for the contents of the container to escape. Then he removed the gauze covering. Three narrow bands of metal ran all the way around the upper edge of the container. He explained to me that an electric current ran through these metal bands at all times. The reason for the electrified barrier was the same as for the watery moat: to ensure that the subjects of this experiment could never escape. Egg cartons were strewn over the bottom of the container—and they were seething with bodies. The whole thing was in motion. Hundreds, no, thousands, of small cockroaches were scuttling around down there. An unpleasant smell rose up to meet my nostrils, a smell that was to accompany me at all times for the duration of the coming year.

"We're researching the aggregation behavior of the German cockroach, *Blattella germanica*," Professor Dambach explained. "Bayer is interested in a substance found in the insects' feces. This substance causes the insects to form clusters. In other words, they aggregate. Therefore: aggregation behavior. Bayer is funding your position."

"Feces?"

"Yes, cockroach excrement," he said, as I stood in the small room where he was breeding cockroaches.

From the rainforest to cockroach poop. That's quite a letdown, I thought. And yet I accepted. My undergraduate thesis was now set: instead of shimmering hummingbirds in South America, I was going to be studying cockroaches in Cologne. My prospects for the coming year were somewhat less glamorous than anticipated. Or were they?

2

The Not-So-German Cockroach

The German cockroach, *Blattella germanica*, is surely one of the least loved animals in the world. When one of these household pests runs right in front of you across the kitchen floor, all the alarm bells in your head go off. For good reason. Experts tell us that for every individual cockroach you see, thousands more are lurking. The German cockroach has followed humanity and spread around the world. This relatively small cockroach, usually less than three-quarters of an inch (two centimeters) long, possesses a couple of characteristics that explain its singular success.

German cockroaches are extremely undemanding. A little warmth, a couple of cracks for shelter, and a crumb or two for food are all it takes for *Blattella germanica* to move in with people. Every hut with a modicum of heat in winter and a small supply of food is a potential habitat for this species.

People have been living in permanent structures for approximately ten thousand to twenty thousand years. At some point, the German cockroach must have thrown in its lot with *Homo sapiens*. What is astonishing is that, as far as we know, *Blattella germanica* no longer exists in the wild. *Blattella germanica* has devoted itself completely to life with us. True, a few thousand years are a mere blink of an eye in evolutionary terms, but it could well be that this cockroach has adapted so completely to life with humans that it can no longer survive without our company. If that is indeed the case, it shares its fate with many other species of animals, plants, and fungi that exist exclusively in indoor ecosystems.

What at first looks like a case of inadvisably extreme specialization is, in fact, an enormously successful strategy. Today, depending on the source you consult, human habitations occupy from 1 percent to 3 percent of all ice-free land worldwide. And while forests, grasslands, and wetlands are shrinking, villages and cities continue to grow up as well as out. Take New York, for example. Manhattan occupies 22.82 square miles (over 59 square kilometers). Because almost all the residential buildings are more than one story high, the potential living space for the German cockroach in Manhattan is three times the area of the island. When you consider the inexorable growth of cities, the future for *Blattella germanica* seems bright.

<hr>

COCKROACHES SPEND MOST of their time in hiding, venturing out at night to look for food. If you come across a cockroach in the light of day, that is not a good sign. In the words of exterminators, you now have an infestation. What that means is that all the roach hiding places are jam-packed with roaches. Cockroaches love company. They huddle together in cracks, using their antennae to maintain contact with the male or female cockroach next to them.

The places where cockroaches hide are not particularly inviting (unless you are a cockroach), because the insects like to do their business where they hide. To put it another way:

they bed down in their own excrement. No wonder we find them nasty to live with. Any creature that has spent the day sleeping in its own feces is unlikely to be welcomed as a guest in the pantry.

Despite its apparently filthy habits, the German cockroach is a fairly harmless housemate. Unlike a bedbug, it doesn't bite, and it doesn't make noise like a house cricket—a kind of cricket that chirps loudly and also likes to move in with humans. *Blattella*, it's true, can trigger allergies and spread diseases, but compared with ticks (which cause Lyme disease and meningitis) or fleas (think of the ravages of the plague), the damage it does is negligible. Nevertheless, the cockroach has become a symbol of neglect. Cockroaches are supposedly a sign of bad domestic hygiene and we suspect that people who have cockroaches don't clean as often as they should. No one wants to have cockroaches skittering across their kitchen floor or have any contact with people who do.

Let's start with the name. Of all the countries one could choose, why are the little pests called German cockroaches? It's not because the species originated in Germany; we still don't know its place of origin. At first, we thought Europe was the place *Blattella germanica* called home. However, most related species of *Blattella* currently living in the wild are found in the tropical jungles of Southeast Asia. We have no idea if *Blattella germanica* originated there as well—so why *German* cockroach? The sobering answer is that *Blattella germanica* was called after the ethnic group that other people wanted to denigrate. The Austrian behavioral scientist Karl von Frisch wrote this about *Blattella*:

In many places in South Germany, they are known as "Prussians", in the North as "Swabians", in West Germany they are called "Frenchmen", in the East "Russians". In

Russia they are again "Prussians"... We cannot help thinking that even scientists have not been without the patriotic prejudices we have just indicated. For it was the famous Swedish naturalist, Carl von Linné, who gave the name... *germanica* to this species.[1]

Carl von Linné, also known as Carl Linnaeus, was a professor at the Uppsala University in Sweden when, in the eighteenth century, he came up with an ingenious idea for naming all species of plants and animals. And thus, we got the system of binomial nomenclature we still use today—every species identifier consists of the genus name followed by the species name. In the case of the German cockroach, *Blattella* is the genus and *germanica* is the species. Binomial nomenclature has unrivaled advantages. It is precise and applies worldwide, transcending all language barriers. Additionally, the name provides information about which group of lifeforms a species belongs to, and thus its position in the plant and animal world. In the case of *Blattella germanica*, a biologist immediately knows that it is an insect in the cockroach family. Why Carl Linnaeus chose to make it a German insect is a secret he took with him to his grave in 1778.

~~~~~~

A MUCH MORE exciting question is how *Blattella germanica* managed to become such a stunningly successful pest, one that can be found in every country in the world. Exterminators call *Blattella* "the perfect bug." Its success begins with cockroach sex—that is to say, the insect's sexual behavior. During mating, the male cockroach transfers a packet of sperm to the female. For several months after mating, the female can produce what are known as oothecae, or egg packets. She carries these around with her until shortly before the fifteen to thirty-five colorless mini-roaches (nymphs) inside hatch. At this point, she takes a long sip of water, which increases the pressure inside her body. The ootheca then falls from the expectant mother's abdomen. A few hours later, her offspring free themselves from their egg packet. In just six to ten weeks, the nymphs undergo multiple larval stages until they become imagos or fully developed adults, which then mate as soon as they can with others of their kind. Theoretically, just five fertilized female cockroaches could—if you include their children and their children's children—produce 45 million offspring in a single year. In reality, however, this number is "only" 100,000 offspring for each pair of cockroaches. Still, a couple of fertilized females stowed away in the provisions of a camel train is all it takes to conquer a new village, a new city, or even a completely new country.

And so *Blattella* conquered the world. Cockroaches have been found in televisions and microwave ovens, in submarines and airplanes. U.S. biologist Eugene Garfield dubbed *Blattella germanica* "the most frequent flyers" of all.[2] If you carry around semen and egg packets, hitch a ride frequently in planes, and produce thousands of offspring, then the world is your oyster.

Cockroaches, however, were on the move long before airplanes were invented. Way back in the sixteenth century, the British privateer Francis Drake reported that he had seized a Spanish galleon that was full of cockroaches. And even today, *Blattella germanica* travels the seven seas in cargo and passenger ships. Its tough, flat body, which fits into the tiniest of cracks, makes it the perfect stowaway.

~~~~~~~

REPRODUCTION AND MOBILITY are not the only things that make cockroaches such feared pests. These insects are not fussy at all about what they eat. Just about anything humans store for food, and any scraps they leave after eating, are more than good enough for the German cockroach. They have specialized cells called mycetocytes. These cells harbor bacteria that help the insects manufacture essential and exceedingly complex compounds from simple organic substances. This symbiotic relationship with bacteria allows cockroaches to survive off a wide variety of unexpected items. For example, twelve cockroaches supposedly survived for a week by eating the glue off a single stamp.[3] *Blattella* have also been observed making a meal of shoe polish, fabric, and human hair.

The self-sufficiency of cockroaches makes it doubly difficult to control them. Once these creepy-crawlies have settled in somewhere, it's not enough to simply say, "Tidy up and clean more." Sometimes, more cleaning can end up having the opposite effect. When I started my thesis work at the Institute of Zoology in Cologne, I was told the following story:

For years, the biologists at the institute had been battling a veritable plague of cockroaches. Anyone who went to their lab at night to check on things couldn't believe what they saw when they turned on the overhead light. The whole floor would be brown, completely covered with thousands of tiny cockroaches. Startled by the light, the insects scooted under lab equipment and squeezed into cracks in the wall. Within a few seconds, the university's linoleum regained its original color: light blue. At night, the insects ate everything they could find—even checking out photo negatives hung up to dry. Precautionary measures were decreed. No sandwich scraps were to be discarded in the wastepaper baskets, nothing edible was to be left lying around. But it was no use. The cockroaches were unstoppable. And the efforts at hygiene had quite the opposite effect—the more vigorously the university's janitors cleaned, the worse the situation became.

The key to the puzzle turned out to be the floor wax the janitors were using. The manufacturer used to mix small amounts of the pesticide DDT into the wax. Then, in the 1970s, this pesticide was banned because it was accumulating in the natural world. Birds with too much of this poison in their blood became infertile and laid eggs with thin shells. A massive die-off of birds seemed inevitable. And so, the pesticide was placed on the list of prohibited substances. Unknowingly, every time the janitors now waxed the floor, they were laying out a feast for *Blattella* and its symbiotic bacteria. Instead of killing off the pesky creatures every time they polished the floor, the cleaners were providing so much food that cockroach numbers exploded. These days, manufacturers of floor wax and furniture polish mix in repellents to discourage insects and spoil their appetite for household cleaners.

This almost invincible pest, an insect that eats floor wax, was now the subject of my thesis. The more I read about the German cockroach, the more fascinated I became. (That didn't mean, however, that I wanted to share my own home with *Blattella*. Every evening I checked carefully to make sure that no six-legged would-be housemate had slipped into my pockets.) It gradually became clear to me that cockroaches were anything but boring. They are not colorful winged acrobats like hummingbirds, to be sure, but they are tough survival artists and perfect traveling companions for humans.

In short, you could say that the cockroach is the perfect bug.

3

~

Higher and Lower Orders of Animals

Which sounds more exciting—cockroach research or hummingbird research? For most people, the answer is clear. They would rather spend time investigating the effervescent birds of the Americas than the much-reviled cockroach.

But why is this? Why are the primate researchers Jane Goodall and Dian Fossey known around the world, while multitudes of other zoologists never hit the headlines in the same way? The fault lies with the subject of their research. It is because people have divided the animal kingdom into separate classes.

Here I would like to call on a comparison from the world of sports. Vertebrates, especially our closest relatives—mammals, and perhaps birds as well—play in the premier leagues. The second-tier league is reserved for reptiles, amphibians, and fish. All the rest, creatures in the diverse group of invertebrates, where species number in the thousands, play in various amateur leagues. These invertebrates—sponges, snails, flatworms, segmented worms, roundworms, hydras, sea anemones, sea pens, and many, many more—are subjects for dedicated enthusiasts, nerds, and other slightly off-kilter individuals. The more an animal resembles us, the more interesting we find it. Great apes therefore play in the championship leagues, while earwigs and earthworms are relegated to the lower divisions of life.

Biologists with their talk of "higher" and "lower" animals have long been to blame for these oversimplistic ratings. Such epithets, it turns out, are erroneous, as the endlessly adaptable kitchen cockroach *Blattella germanica* attests.

~~~~~~

CERTAIN ANIMALS ARE clearly somewhat simpler than others. That is to say, they are less complex, whether it be in their body structure or in their behavior. For some, however, the illusion disappears if you take a closer look. Then you find absolutely amazing adaptations for the great struggle for survival even in what are supposedly the least complicated of creatures.

Jellyfish, for example, have been swimming in the world's oceans for more than 500 million years. Mostly they drift on the oceans' currents. They can, however, actively propel themselves forward by quickly contracting their bell so water squirts out behind them. The contraction catapults the jelly a short distance, and some species of jellyfish can achieve speeds of over six miles (up to ten kilometers) per hour using this method.

The structure of a jellyfish is not particularly complex. The creature consists of a bell-shaped body and tentacles, which it uses to get food. Jellies have a stomach and a mouth, but no anus. Then they have a few nerve cells and reproductive cells— and that's about, but not quite, all. Anyone who has come into contact with a jellyfish will be well aware that, for all their simplicity, they are equipped with dangerous weapons capable of inflicting great pain.

Jellyfish tentacles are covered in specialized stinging cells called cnidocytes. These high-speed, high-performance

weapons are among the most complex cells to be found anywhere in the animal kingdom. It is not an exact comparison,
but a jellyfish with its deadly stinging cells is a bit like an
ancient Viking ship armed with the most technologically
advanced laser-guided missiles.

These stinging cells have just one function, but it is an essential one if the jellyfish is to survive. The cells must inject, in the
most effective manner possible, a dose of venom into the bodies of other creatures, whether those creatures are predators
that want to eat the jellyfish or prey the jellyfish wants to ingest.

The jellyfish has a couple of problems: it is not particularly
fast and it doesn't have sharp fangs to inject venom into its
victim as, say, a cobra does. Instead the jellyfish uses its stinging cells. Inside these cells, there is a kind of coiled tube that
contains venom. If the tentacle of a jellyfish, along with its
stinging cells, brushes up against an unfortunate swimmer or
an even more unfortunate fish, tiny sensors on the outside of
the cells register the contact. But a chemical stimulus is also
needed to set one of the animal kingdom's deadliest processes
into motion. The cells taste what they have touched. They can
discern whether it is another of the jellyfish's tentacles, a piece
of driftwood, a predator, or a little fish that is suitable prey.

Then everything happens in a flash. The stinging cells literally explode and some launch harpoon-like structures that
accelerate so quickly they can penetrate the shells of crustaceans and the scales of fish. The stinging tubes now protrude
like the fingers of a rubber glove, and the jellyfish's venom is
discharged into the victim. Barbs on the many stinging tubes
ejected in a strike ensure that the prey is secured and cannot
swim away before the jellyfish venom kicks in.

Biologists Thomas Holstein and Pierre Tardent[4] filmed this
process, known as exocytosis, with ultra-high-speed cameras
that take many thousands of frames per second. The footage

showed that only three one-thousandths of a second pass between first contact and the venom being successfully delivered to the victim. The stingers are ejected at an unbelievable top speed of 4,350 miles (7,000 kilometers) per hour, which means they are launched with more explosive force than a bullet from the barrel of a hunting rifle.

The stinging cells' launch mechanism is not the only part of the jellyfish's weaponry that is unusual: the venom of some jellyfish is a wonder of nature. The venom in the tentacles of tropical box jellyfish is one of the most dangerous of all for humans. Entire beaches in Australia are fenced off on the ocean side to protect swimmers from the box jellyfish *Chironex fleckeri*, which goes by the common name "sea wasp" and is capable of stinging an adult to death.

Should an animal armed with such high-tech weaponry really be relegated to the lower leagues? I don't think so. In the end, what counts is whether a life-form is successful or whether it dies out. Thanks to their super-effective defensive cells, jellyfish and other sea creatures with stinging cells, such as freshwater polyps, corals, and sea anemones, have been able to survive for several hundred million years without fundamentally changing their shape, despite their relatively simple physique. They have no need for fangs, no need for muscles that power legs or wings to move quickly, nor any need for a brilliant brain—an animal that has such effective stinging cells can survive for millions of years.

~~~~~~

DESPITE THEIR ABILITY to put up a fight, jellyfish and their ilk do have animals that hunt them. Some of them—a few species of nudibranchs and comb jellies—consume all parts of their venomous victims except for the stinging cells. They separate out these tiny weapons and store them on their own skin, thus making themselves untouchable, like the German hero Sieg-fried after he bathed in dragon's blood or the Greek warrior Achilles after his mother dipped him in the river Styx. Biol-ogists call this kleptocnidy, from the Greek word *klepto*, "to steal," and the Latin word *cnide*, or "nettle." Stinging cells are so coveted, complex, and precious that animals become pred-atory killers in order to possess them.

A small species of crab, the boxer crab, *Lybia tessellata*, has found another way to use sea creatures with stinging cells in its own defense. It cultivates a venomous sea anemone on each of its two claws. When the little crab threatens an enemy with its stinging tenants, it looks like a boxer sparring for a fight—and that's how the little boxer crab came by its name.

One could certainly argue that jellyfish, polyps, corals, and sea anemones might be an extremely successful group of ani-mals thanks to their stinging cells, but they can exist only in water. Sweeping evolutionary advances, plus complex organs and behaviors, are necessary if animals are to conquer new habitats. Without specialized breathing apparatuses, there would be no bison roaming the plains; without circulatory sys-tems and temperature regulation, no polar bears patrolling the Arctic; without wings and muscles, no hawks soaring on updrafts. All that is true. However, the so-called higher vertebrates are not the only animals to have mastered these evolutionary advances.

Long before amphibians crept out onto land, the ancestors of today's arthropods—primordial crabs, centipedes, millipedes, spiders, scorpions, and insects—all left the ocean. Even such supposedly simple life-forms as segmented worms, slugs, and snails took leave of the watery depths.

Before there were pterosaurs, bats, or birds, insects devel-oped wings and flew. Today, about eighteen thousand species of birds and one thousand species of bats live on Earth, while most of the million (or more) species of flying insects don't even have a scientific name.

So why is it that we divide animals into higher and lower orders? This hierarchy was imposed by biologists such as Ernst Haeckel, who was born in 1834 in Potsdam, in what is now Germany. Haeckel, a champion of the theory of evolution and a gifted illustrator, used the image of a family tree to explain the novel concept of the origin of species to his contemporaries. And, naturally, who was enthroned right at the top of this tree of life? Humankind, of course, with the great apes just below and prosimians (lemurs, lorises, bush babies, and tarsiers), cats, dogs, and other mammals at the periphery of the crown.

A little farther down the trunk, birds, reptiles, amphibians, and other vertebrates branched off. Then there were the arthropods and mollusks, and even farther down, barely above the ground, the stinging sea creatures, the cnidarians. The problem with Haeckel's tree of life is its hierarchy: up above sits the crowning glory of creation and below lies the dross. The tree of life gives the impression that evolution moves in only one direction and has a single purpose: to produce highly complex humans superior to all other animals.

Today, evolutionary biologists are attempting to replace the tree of life with a more fitting image—a kind of multistemmed and densely branched bush. Some of its stems are thicker than others, and they keep dividing. In this image, all species of animals alive today can be found at the tips of the many branches of the bush. The distance between the branch forks and the exterior of the bush illustrates how far back in Earth's history the split happened, and therefore how long it has been since this group of animals arose. In this image, there are no higher or lower orders of animals, no animals that are better or worse than others. The only feature of importance is which life-forms persisted and which did not.

THE EVOLUTION OF humans is a wonder, no question. But it is just one of many wonders. What's really amazing is the enormous variety of life expressed through millions of different species. These include the sperm whale, which dives up to six thousand feet (two thousand meters) deep, as well as well-armed jellyfish and dangerously venomous cone snails that prey on nimble fish, giant trees hundreds of years old, and even six-legged cockroaches, which conquered the Earth by following humans and feel most comfortable bedded down on their own excrement. Every creature is a wonder. Even *Blattella germanica*.

4

Renting— and Gilding— Cockroaches

A nd so, the topic of my undergraduate thesis was to be the aggregation behavior of the German cockroach. The scientific question I was to answer was, of course, somewhat more specific. I was to find out which substance induces cockroaches to make themselves comfortable on their own excrement. Just a few days later, I was cutting blotting paper into narrow strips, which I sprinkled with an extract made from cockroach feces. A craft project in the name of science...

Many animals communicate with others of their own kind using scent: they produce substances called pheromones, which they release into their environment to exchange information with one another. When a male dog marks his territory, he's telling other pooches in the neighborhood that this is his turf, so they'd better watch out. Or a female dog could be advertising that she is in heat. But pheromones don't only communicate information; they also alter the behavior of the animal that picks up the message. Sticking with the example of the male dog: assuming his owner does not stand in his way, he will track down the receptive female as quickly as he can.

For most insects, the way this communication works is very simple. When they come across a message from one of their own kind, their behavior changes as if a switch has been activated. It happens with *Blattella* too. When the antennae of cockroach number one come into contact with the droppings left by fellow cockroach number two, cockroach number one settles down on top of the pile. As cockroach number one then emits its own scent mark, roaches three and four join the group. As the messages pile up, what started as a small group expands to form a colony of commanding proportions. These cockroach clusters disband at night as the creepy critters go off in search of food. As the sun rises once again, they return, and contact with the cockroach droppings puts the insects back into sleep mode. But which of the many substances in the droppings has such a calming effect on the cockroaches? The hope was that if I found the pheromone that triggered aggregation, exterminators would have an effective weapon to use in their daily battles against these little pests. A pheromone-based cockroach trap could turn into a lucrative business proposition. And that was why the corporate giant Bayer was funding my modest thesis project.

What followed was teamwork. Bayer chemists distilled one extract after another from cockroach feces. I tested the many trial doses using a behavioral test. My "aggregation test" was ingeniously simple. I folded two strips of blotting paper so each formed a right angle. Then I sprinkled one with the extract to be tested and the other, the control, with a solvent. Then I put the two strips down to create a small test arena. Finally, I herded ten freshly hatched nymphs (baby cockroaches) into the arena: my laboratory animals.

It didn't take long for the youngsters to gather on one of the two strips. One fecal extract had a 100 percent success rate in attracting the insects to the test strip while none went to the control strip. Bayer chemists now wanted to continue refining this *Blattella* fecal extract into smaller and smaller component parts. The test extracts that performed well in the aggregation test would be investigated further and, after more tests, those that proved effective in the second round would be broken down even more. They wanted to keep testing and refining the extracts until one was so pure it contained nothing but the active ingredient: the aggregation pheromone they were looking for, the exterminator's holy grail.

~~~~~~

WHILE I WAS busily raising roaches and funneling the next generation of nymphs into my test arena, I told friends and acquaintances that my research had "something to do with cockroaches." It wasn't long before I got the first call.

"Frank, my garbage can is full of maggots, you HAVE to help me!"

"Hey, what can I do about flour moths?"

"Help! There are slugs crawling all over my garden!"

Once they learned I was studying the behavior of cockroaches, my friends immediately assumed I was majoring in pest control. I turned down nearly all the appeals for help. But then a message on my answering machine piqued my interest.

"Say, I hear you're raising these little guys up there at the university. Could you rent me a couple for a film I'm making? The gig pays well, of course!"

The call came from Ralph, a friend who worked as an editor for VIVA, a music station that was just getting off the ground. I called back immediately. The idea of being the first person in the world to establish a successful business renting cockroaches was just too tempting.

My editor friend explained that they had an idea for a film. The plan was to put cockroaches in a dollhouse and then use a voice-over to have them say silly things.

"We're going to call the program *The Roaches*," he explained. "You know, with laugh tracks like in American comedy shows."

Yes, I thought. This could be really silly.

I quickly agreed, but just as quickly I had second thoughts. What if an entire television studio got overrun with a plague of cockroaches because of me? I took a moment to imagine what that might look like. It was not a pretty picture.

Then, however, I got a lucky break. New residents moved into the cockroach breeding room: giant Madagascar hissing cockroaches. These insects are as impressive as their name suggests. The jet-black crawlers grow up to 2.75 inches (7 centimeters) long—about the length of a human index finger. When they feel threatened, they hiss by expelling air from breathing holes in their abdomen. They sound like angry cats. These insects, I decided, would be the stars of the VIVA television series. They had a number of things going for them. Because Madagascar hissing cockroaches are so well armored, they do not need to scamper away quickly. Therefore, the studio would not be in danger. Their gigantic size was another point in their favor. The camera operators would not need any expensive macro lenses to film them, and the Madagascar cockroaches had a more commanding presence than *Blattella* inside the dollhouse that was to serve as the backdrop for *The Roaches*. Giant cockroaches are not particularly attractive, but that was the whole idea.

A few days later, Ralph and I met conspiratorially in the Institute of Zoology parking lot. I handed over twenty Madagascar cockroaches to the trendy editor in return for two hundred German marks in cash. Not a bad return for five minutes' work. I advised my friend to pop the armor-plated actors into the fridge for a bit just before filming.

"Lightly chilled, they are less mobile and less apt to hiss. Your team will thank you."

~~~~~~

I SOMETIMES WONDER what was worse: that I allowed the filmmakers to tie the lightly chilled insects to dollhouse furniture with sewing thread—and then film them after they had warmed up, and were waving their six legs around—or that the gag writer put words into their mouths like this (in the dialect common to Cologne, no less):

"We should hit the dance floor again. We'd cut a mean rug out there." (Laughter from the band.)

"We can't do that."

"Why not?"

"There is no rug! The dance floor's made of wood!" (More laughter from the band.)

The Roaches—Let's Go Dancing! is the first nature film I was involved with. This snappily written opus ran for a grand total of twenty-seven seconds. My friend was delighted and further episodes of *The Roaches* were produced. All the insect actors survived filming without so much as a bent antenna, and with every cockroach delivery, I augmented my modest undergraduate salary. Everyone knows students are perennially short of money.

~~~~~~

MEANWHILE, MY RESEARCH continued. Every week, more packages arrived from Bayer with new test samples. A few worked like a charm. They attracted the baby cockroaches to the test strips and were researched further. Others had no effect at all and were discarded. However, something was happening that worried everyone involved in the experiment. The more the Bayer lab refined the cockroach extract, the less effective the individual samples became. The hope was that the aggregation pheromone would end up in concentrated form in one of the samples. This extract would then function fantastically well, and the others wouldn't work at all. But we had the pheromones all wrong. Maybe what the cockroaches were drawn to was the whole cocktail, the mix of many different substances extruded from the cockroach gut along with the droppings. What we needed to find out was whether the common cockroach, *Blattella germanica*, actively added a chemical message to their calling cards. Was there a gland that produced the elusive pheromone and inoculated the feces with the miracle product we were seeking?

~~~~~~

I ADMIT THE answers to these questions were not going to help humanity progress or make our world a better place. It was also absolutely clear to me that this research wasn't going to help me solve some great evolutionary puzzle. Nevertheless, I decided to expand my research into an area heretofore undeservedly neglected by the international scientific community: the lower gut of *Blattella germanica*.

My method, I decided, would be to examine the cockroaches under a scanning electron microscope—SEM, for short. In an SEM, a high-energy electron beam is directed at the object to be magnified. When this primary beam reaches the object, it spins other electrons out from the surface of said object. These secondary electrons are directed to a focusing screen, where they create a greatly magnified image of the object under investigation. In my case: cockroach rear ends. But if you are going to take a closer look at an insect under an SEM, it must first be coated with a fine layer of a highly conductive metal. That is how you make sure the electron beam doesn't super-charge the test object and everything doesn't end up going up in flames—including the shockingly expensive scanning electron microscope.

Gold is one such highly conductive metal. And yes, in the course of my thesis work, I did gild the rear ends of cock-roaches. To be more precise, I steamed them lightly in gold in a vacuum chamber—after they were already dead, of course. Then I slid them into the SEM.

When I rented out cockroaches, I supplemented my income. When I looked at them under a scanning electron microscope, I entered a completely new world.

5

Devoted Parents

The monitor of the high-tech machine—the scanning electron microscope—revealed wonders hidden to the naked eye. Magnified thousands of times, *Blattella* became a creature reminiscent of the alien in the film of that name. Fine hairs covered the smooth chitin casing. A particularly large number of these hairs bristled from the cerci, two antenna-like feelers protruding from the insect's abdomen. The moment the cockroach feels the slightest breath of air passing over these hairs, its six cockroach legs immediately begin to move. This flight reflex is a lifesaver for an insect that has neither sharp spines nor sharp teeth to defend itself.

I, of course, was most interested in looking at the cockroach's bottom. There, I could make out small bulges known as rectal papillae. Insects use them to extract precious water from their droppings, an adaptation to living in dry regions of the world. *Blattella* has tiny hooks on its rear end that it uses to hold on to its partner during mating. There were protuberances next to them, and I could make out minuscule pores on their surface. The diameter of these openings was less than one-thousandth of a millimeter. Perhaps these pores connected to the pheromone gland we were looking for? Unfortunately, I could not prove that; however, the view under the SEM was still invaluable.

The longer I looked, the more fascinated I was by the tiny insect. Unremarkable-looking hairs turned out to be highly sensitive motion detectors, and I could make out incredibly delicate veins in the cockroach's wings. The cockroach's compound eye came into focus as a complex work of art in which countless spherical lenses were combined in an orderly array. The scanning electron microscope irrevocably changed my relationship with insects. It was love at second sight.

Despite my infatuation, I had to refocus my attention on fecal matter. Perhaps cockroach excrement is attractive to *Blattella* for other reasons beyond being a carrier of an as-yet-undiscovered pheromone. The next order of business was to check out close relatives of the German cockroach. Were there other species of cockroach that also had a special relationship with their own excrement? What started as a literature review on the subject of cockroach droppings revealed yet more wonders. Cockroaches are not simply pests. Many species are highly social. Some maintain strong family bonds. And their droppings could well have played a not-insignificant role in the development of their social behavior.

～～～～～

ROACHES IN THE genus *Cryptocercus* are the real helicopter parents of the cockroach world. All members of this group share a taste for an unusual food: they eat the wood of dead tree trunks. *Cryptocercus* cockroaches gnaw their way deeper and deeper into downed trees. With every meal, they increase the length of the tunnel they are excavating and create a safe home where they can live. Wood, however, isn't particularly nutritious. There are only a few life-forms that can digest wood, and most of them, including *Cryptocercus*, need the help of specialized microorganisms. Just as humans have gut bacteria, so, too, *Cryptocercus* cultivate bacteria and single-celled organisms in their gut—in their case, ones that help them digest wood. They have a problem, however, because *Cryptocercus* nymphs hatch from their eggs minus their complement of symbiotic helpers. Left alone in their woody homes, *Cryptocercus* babies would starve. Nature, of course, has a handy solution to this problem.

Cryptocercus cockroaches are devoted parents. Mother, father, and offspring form small family groups that share a home for many years. During this time, the young larvae are fed by their parents. As heartwarming as this sounds, I should mention that the cockroaches' parental nurturing doesn't happen from mouth to mouth. Instead, the offspring feed on small particles of wood that the parents excrete in their feces. These first few weeks of eating parental droppings don't just help the little ones survive. As they ingest their unusual baby food, they are also ingesting the symbionts they need, and with the help of these microorganisms, the nymphs can move on to eating wood themselves. It takes more than five years for *Cryptocercus* nymphs to become sexually mature—five long years that they spend with their parents, like spoiled teenagers. Only then do they leave the family's living and eating quarters to seek their own dead trunks and start their own small family units.

~~~~~~

COULD SOMETHING LIKE this be going on with the German cockroach? Did its aggregation behavior develop in a similar way? Did the ancestors of *Blattella* perhaps feed their offspring with their own droppings? Maybe *Cryptocercus* are not the only cockroaches that deliver vital symbionts in the droppings that fall from their rear ends. Might that be why it is so important for German cockroaches in all stages of life to gather on their feces every night? I admit that the whole subject of eating droppings is not particularly inviting, but it might explain the origin of social and parental behavior in some insects.

Other species of cockroaches look after their young just as attentively as *Cryptocercus*. Females in the species *Diploptera punctata* give birth to fully developed nymphs. Just as mammal embryos develop in the womb, these cockroach young develop in a special brood pouch in the mother. The expectant mother feeds her unborn nymphs with a substance secreted by a gland in her brood pouch. The Indian biochemist Sanchari Banerjee analyzed this cockroach concoction and the results took everyone by surprise. Not only does it contain three times as much protein as cow's milk, but it also contains everything else a cockroach needs to grow: fat, carbohydrates, and trace elements. In short, *Diploptera punctata* feeds its offspring with a mother's milk specially formulated for cockroaches.[5]

Cockroach mother love can go even further. After hatching, nymphs in the Indian species *Thorax porcellana* snuggle up under their mother's wings. About fifty baby cockroaches at a time find not only protection, but also very special baby food. *Thorax* parental feeding more closely resembles a scene from a vampire film than a romantic family story. The young are born with razor-sharp mouthparts. They use these to bite holes in their mother's chitin covering, and for the next seven weeks they lick up her hemolymph, a kind of insect blood.

~~~~~~

"COCKROACHES ARE MORE diverse in their reproductive biology than probably any other group of insects," British entomologist George Beccaloni told me. For thirteen years, he was the curator of the collections of cockroaches and their relatives at the Natural History Museum in London. While on vacation in England, I paid a visit to the world-renowned museum. You could describe the Natural History Museum as a cathedral to the natural world. This enormous Romanesque temple to science houses some of the most important geological and biological collections in the world. Plants that the botanist Joseph Banks brought back with him from his voyage in the South Seas with Captain James Cook are stored here, along with specimens from Charles Darwin's expedition to the Galápagos Islands and Alfred Russel Wallace's scientific explorations in the Amazon and Malay Archipelago.

There are few people around the world who know as much about cockroaches as Beccaloni, who was kind enough to grant me some of his precious time. As a child, he lived with his parents in what was then Rhodesia and is now known as Zimbabwe. "When I was young, I collected natural history objects ranging from minerals to shells and skulls," the cockroach curator told me. "I was particularly fascinated by the five hundred species of butterflies found in the country." When he was fourteen years old, his family moved to the U.K. That was somewhat of a shock for young George. There were only fifty-nine species of butterfly on these chilly islands, and thousands of amateur butterfly sleuths and hundreds of specialized books about Britain's butterfly populations.

Beccaloni quickly realized that British butterflies were overstudied. If he wanted to become a serious researcher, he would have to find another passion. As people from Britain are wont to do, Beccaloni joined a club, the Blattodea Culture Group. Cockroaches belong to the order Blattodea, hence the name. The combination of Blattodea and Culture suggests its members might have been interested in the cultural history of cockroaches, but that was not the case. Nor were they interested in discussing the importance of cockroaches in opera, film, or literature. What the members were interested in was exchanging information about the best ways to raise—in other words, cultivate—exotic cockroaches in terrariums. The young Beccaloni quickly found a new hobby, one that led to a new passion, employment, and, in the end, a career.

Later on, Beccaloni created the Cockroach Species File, an official catalog of cockroach species that, at the time of writing this book, contained 4,600 species. The British cockroach expert estimates there are about another 20,000 species that do not yet have scientific names. Most of these as-yet-undescribed species are crawling around in tropical rainforests in South America, Africa, and Asia. Beccaloni defends his favorite animals by saying, "4,600 *known* species. That is almost exactly the number of mammal species worldwide. Of the thousands of cockroach species, only about thirty are troublesome household pests, so one shouldn't condemn the whole order." We don't do that with mammals, after all—even though mice move into our homes, moles wreak havoc in our gardens, and rabbits munch their way through our vegetable beds. Despite that, we love elephants, protect tigers, and rescue stranded sperm whales. If we are being consistent, then shouldn't we also be concerned for the welfare of all 20,000 species of cockroach?

~~~~~~

THERE ARE SKY-BLUE, yellow, and green cockroaches. Others are bright red with black spots. If that last one sounds familiar, you are on the right track. These cockroaches are imitating nasty-tasting ladybugs so they themselves don't get eaten.

It took more than seventy years for a cockroach found in Ecuador in 1939 to hit the headlines. For decades, the desiccated insect was stored in the Smithsonian National Museum of Natural History in Washington, D.C. Then one day, Slovakian cockroach researcher Peter Vršanský took a closer look at the specimen. Two vibrant yellow spots were emblazoned on its back. Vršanský concluded that these spots were light-emitting organs. He speculated that this cockroach, a member of a species that had not yet been identified, glowed at night. The cockroach, so Vršanský's theory went, appeared to be imitating large tropical fireflies in the click beetle family. The lights of click beetles are a warning to insect-eating bats: "Watch out! I taste bad." Vršanský theorized that the glowing cockroaches hijacked this message. He speculated that because insect-eating bats mistook the light-emitting cockroaches for bitter-tasting click beetles, they gave them a wide berth.

We will likely never prove whether Vršanský was right. Rarely documented, the glowing cockroach could well be extinct by now. *Lucihormetica luckae*—that's its scientific name—had been discovered in a tropical cloud forest in Ecuador seventy years earlier on the slopes of the 16,480-foot-high (5,023-meter) Andean volcano Tungurahua. Over the centuries, humans have destroyed most of these tropical cloud forests to make way for roads, cow pastures, and cities.

Unfortunately, Tungurahua is also one of the most dangerous and most active volcanoes in the world. Every couple of decades it spews out burning chunks of rock and mile-high ash clouds. Hot lava flows from its crater and down the mountain. People living in the nearby town of Baños de Agua Santa have to get out of harm's way. A whole cloud forest complete with its inhabitants, however, cannot be evacuated. Tungurahua's frequent eruptions destroyed what little forest was left—and with it, probably the only known population of the glowing cockroach *Lucihormetica luckae*.

~~~~~~

DESPITE THE DIVERSITY in appearance and behavior of the many cockroach species, scarcely any have developed really dangerous defensive weapons. There are no venomous, biting, or stinging cockroaches. A few species can spray an unpleasant-smelling secretion, and giant Madagascar hissing cockroaches like the ones I rented out sound the alarm when they feel threatened, but that's about it for active defense mechanisms. Camouflage, concealment, and running away are the most important tools in the cockroach survival kit. Given how harmless they are, I never cease to be surprised at how disgusting many people find them.

Cockroaches live on all continents except Antarctica—although someone would need to confirm whether *Blattella* has not perhaps been resident in one of the many Antarctic research stations. Cockroaches browse through the foliage of rainforests and crawl through grasslands and deserts. *Eupolyphaga everestiana*, as the name implies, survives on Mount Everest at elevations of up to 18,500 feet (5,600 meters).

"Other cockroach species have adapted to highly specific habitats," Beccaloni explained. He proudly told me about the diversity among the many species of cockroaches, including a species that has taken to the aquatic life. Bromeliads are plants that grow in the tropical rainforests of South and Central America. Their thick leaves form a cup where rainwater gathers. Mosquito larvae and tadpoles live there, as do some species of frogs. And there are cockroaches that specialize in life in the bromeliad pools. They bravely leap into the mini water bodies and power to the bottom using their three pairs of legs. There, the armored divers hide from predators, living off the plant debris and dead animals that accumulate in the bromeliad pools.

Each of the thousands of species of cockroach on this planet has a different survival strategy. Beccaloni has observed untold numbers of them, but sometimes even he is surprised. When danger threatens, Southeast Asian cockroaches in the genus *Perisphaerus* roll themselves up like armadillos. That on its own is nothing extraordinary; however, another cockroach researcher was once examining a dead female in a museum collection and he couldn't believe his eyes. Tiny nymphs were clinging onto her legs. They had died along with their mother in the name of science.

The researcher plucked a dead nymph from one of the female's legs. The mouthparts were completely different from those of the adult. Normally, cockroaches have powerful jaws so they can carve their food up into bite-sized pieces. *Perisphaerus* nymphs have a delicate drinking straw on their face instead. This straw is slightly smaller in diameter than four strange holes between the mother's legs. Since *Perisphaerus* nymphs are blind and transparent, they would not survive for long in the wild without their mother's care. And they don't have to, because she carries them with her at all times for the first weeks of their lives. The little ones don't have to do much other than suck up a nutritious secretion their mother delivers via the holes—yet another cockroach mother that nurses her offspring with cockroach milk.

Biologists refer to it as "analogous" when distantly related organisms independently develop similar organs or behaviors. The parental care of people and the parental care of *Perisphaerus* are strikingly similar. In biological terminology, the fact that two such dissimilar animals carry helpless offspring around and raise them on nutritious milk is astoundingly analogous behavior. Could this similarity—the fact that cockroaches, like us, are social, care for their children, form family units, compete with us for our food, and make themselves at home in our houses—be part of the reason we are so wary of them? Do we find them, in the end, simply too much like us?

Before I left the museum, I took advantage of my conversation with Beccaloni to ask the ultimate cockroach question. Is it really true that cockroaches can survive a nuclear explosion?

Did they survive Hiroshima, as people constantly claim? Finally, will they survive us if human civilization is wiped out in a nuclear war? "That's one of those urban myths," Beccaloni said dismissively. "The heat of a nuclear explosion would destroy insects just as it would destroy us." There is, however, a kernel of truth to the myth. Cockroaches deal with radioactivity better than humans, and mammals in general. Experiments have shown that they are about ten times as resistant to radiation as we are. But cockroaches are not alone in this. Radiation is less of a problem for all sorts of creatures, from fruit flies to worms to amoebas. In a world contaminated with radiation, there would be no people, dogs, mice, or sloths. But in this apocalyptic scenario, there would still be cockroaches. And they would be sharing the planet with many other life-forms.

6

In the Footsteps of Fabre

The calls for help continued—and continue to this day. Friends are constantly calling because they are bothered or feel threatened by pesky insects or other creepy creatures. As I mentioned earlier, I usually suggest my acquaintances consult a specialist. There was one time, however, when I made an exception.

"Spiders, maggots, and wasps are living it up in the roll of parchment paper in my baking drawer," my friend Susanne messaged.

She attached a photo as proof, and things did look somewhat chaotic. I could make out fat white maggots, strange brown bumps attached to the parchment paper that looked as though something was living in them, and lots of tiny spiders. The scene put me into detective mode. Spiders never look like maggots and I could think of nothing that connected the eight-legged predators to the brown bumps, which were clearly constructed of soil. What sort of insect revelry was going on in downtown Cologne? I wanted to find out what was happening and help my friend.

I called her up and got more information. I discovered that it had all started with an increasingly loud humming in a kitchen drawer. My friend tracked the sound down to her roll of parchment paper. But there was more. She watched as a large wasp left its hiding spot and buzzed out of her open kitchen window to freedom. She then made the strange discovery in the parchment paper.

Wasps and spiders are predators. But who was eating whom in the baking drawer? One thing was clear: none of the protagonists were showing any interest in Susanne's stores of food. I could sound the all clear on that, at least. The many spiders in the photo looked dead, but they also looked quite fresh. I was intrigued. Not the slightest sign of mold or decay. Did they just look like they were dead? Was my friend dealing with zombie spiders? With that thought, it all fell into place—and I remembered the greatest entomologist of all time, the French scientist Jean-Henri Fabre.

Fabre was a physics teacher and amateur entomologist who lived at the end of the nineteenth century in Corsica and in the South of France. He was not only a good teacher but also a gifted writer. He wrote textbooks that sold so well he could live his dream. Fabre bought a small estate in Sérignan-du-Comtat, not far from the town of Orange. *Harmas* was the name Fabre gave to his new home in the sun-drenched landscape of southern France. For the rest of his life, Fabre pursued his two greatest passions in the environs of Harmas. By day, he wandered through nature and his garden, studying everything that crept and flew. In the evening, he sat at his now-famous desk and put his observations down on paper. That was how he wrote *Souvenirs entomologiques*, a collection of entomological observations. *Souvenirs entomologiques* was far from being a dry text for specialists. It was a literary delight, a mix of precise observations of insect behavior, poems, philosophical thoughts, and journal-type entries. At Harmas, Fabre wrote such gripping stories about nature that he was nominated for a Nobel Prize—in literature.

But what did a French naturalist and author have to do with my friend Susanne's wasp and spider problem?

~~~~~

JEAN-HENRI FABRE WAS one of the first people ever to collect butterflies, beetles, and wasps for something other than their beauty. He was much more interested in how they lived, what they ate, and how they reproduced. Nothing escaped Fabre's curiosity. This made the physics teacher one of the first and most important animal ethologists—more than one hundred years before Konrad Lorenz, Nikolaas Tinbergen, and Karl von Frisch, who up until now have been the only animal ethologists to win the Nobel Prize in Medicine.

Fabre's masterpieces include his observations of the yellow-winged digger wasp. Since then, biologists have discovered and described over ten thousand species of digger wasps, from tiny fliers less than half the size of your little fingernail to giants equipped with fearsome stingers whose bodies are as long as a couple of your fingers are wide. Unlike the common yellow-jackets that spoil our late-summer picnics, digger wasps do not live in large colonies. They live out their insect lives all alone. Fabre's object of study locates an area of loosely packed soil and uses its six legs to dig a hole. It creates a little spray of dirt as it kicks grains of sand and particles of soil up behind it. It works its way deeper and deeper into the ground until its tiny burrow is finally ready.

Fabre observed how the hardworking wasp then flies away, only to return a short time later with a prey animal, a cricket.

It drags the cricket into its burrow, lays an egg on it, and then leaves, closing the entrance behind it.

Thanks to Fabre, we know a great deal more about the amazing reproductive strategies of the digger wasp. These predatory insects are incredibly picky about what they bring into the natal den as food for their offspring. Fabre's yellow-winged digger wasps dragged in almost nothing but crickets from around the den. They focused so exclusively on crickets that the naturalist was quite surprised when he saw a wasp make an exception one day.

> The scene is enacted on a towing-path along the Rhône. On one side is the mighty stream, with its roaring waters; on the other is a thick hedge of osiers, willows, and reeds; between the two runs a narrow walk, with a carpet of fine sand. A Yellow-winged Sphex appears, hopping along, dragging her prey. What do I see! The prey is not a Cricket, but a common Acridian, a Locust! And yet the Wasp is really the Sphex with whom I am so familiar, the Yellow-winged Sphex, the keen Cricket-huntress. I can hardly believe the evidence of my own eyes.
>
> The burrow is not far off: the insect enters it and stores away the booty. I sit down, determined to wait for a new expedition, to wait hours if necessary, so that I may see if the extraordinary capture is repeated. My sitting attitude makes me take up the whole width of the path.[6]

It's no surprise that back in the nineteenth century a grown man sitting for hours in the middle of a path to observe tiny insects going about their daily business was ripe for the suspicion and ridicule of his contemporaries. And that explains why this particular digger-wasp expedition met with disaster.

Two raw conscripts heave in sight, their hair newly cut, wearing that inimitable automaton look which the first days of barrack-life bestow. They are chatting together, talking no doubt of home and the girl they left behind them; and each is innocently whittling a willow-switch with his knife. I am seized with a sudden apprehension. Ah, it is no easy matter to experiment on the public road, where, when the long-awaited event occurs at last, the arrival of a wayfarer is likely to disturb or ruin opportunities that may never return! I rise, anxiously, to make way for the conscripts; I stand back in the osier-bed and leave the narrow passage free. To do more would have been unwise. To say, "Don't go this way, my good lads," would have made bad worse. They would have suspected some trap hidden under the sand, giving rise to questions to which no reply that I could have made would have sounded satisfactory. Besides, my request would have turned those idlers into lookers-on, very embarrassing company in such studies. I therefore got up without speaking and trusted to my lucky star. Alas and alack, my star betrayed me: the heavy regulation boot came straight down upon the ceiling of the Sphex! A shudder ran through me as though I myself had received the impress of the hobnailed sole.[7]

~~~~~~~

FABRE, HOWEVER, DID not let that put a damper on his expeditions and he went on to find out a lot more about his digger wasps. Like bees, the adults visit flowers and feed exclusively on nectar and pollen. Only the females become dangerous hunters. They attack other insects and spiders to supply their brood with protein-rich food. Some species of digger wasps seize nothing but cockroaches; others specialize in crickets. Then there are those that hunt bees, and yet others that tackle an opponent most insects are careful to avoid—spiders.

With Fabre's help, I was getting closer to unraveling the parchment-paper mystery. It was just a matter of finding out which species of wasps focus on collecting nothing but little spiders.

Fabre himself had by no means unraveled all the secrets of digger wasps. He suspected all was not well with the apparently dead crickets his experimental subjects brought to their burrows. Scientist that he was, he stole a motionless cricket from one of the wasps, replacing it with a living one that he had previously caught. Fabre was stunned by what happened next:

Nothing would induce me to give up my part in the tragic spectacle which I am about to witness. The terrified Cricket takes to flight, hopping as fast as he can; the Sphex pursues him hot-foot, reaches him, rushes upon him. There follows, amid the dust, a confused encounter, wherein each champion, now victor, now vanquished, by turns is at the top or at the bottom. Success, for a moment undecided, at last crowns the aggressor's efforts. Despite his vigorous kicks, despite the snaps of his pincer-like mandibles, the Cricket is laid low and stretched upon his back.

The murderess soon makes her arrangements. She places herself belly to belly with her adversary, but in the opposite direction, grasps one of the threads at the tip of the Cricket's abdomen with her mandibles and masters with her fore-legs the convulsive efforts of his thick hinder thighs. At the same time, her middle-legs hug the heaving sides of the beaten insect; and her hind-legs, pressing like two levers on the front of the head, force the joint of the neck to open wide. The Sphex then curves her abdomen vertically, so as to offer only an unattackable convex surface to the Cricket's mandibles; and we see, not without emotion, its poisoned lancet drive once into the victim's neck, next into the joint of the front two segments of the thorax, and lastly towards the abdomen. In less time than it takes to relate, the murder is consummated; and the Sphex, after adjusting the disorder of her toilet, makes ready to haul home the victim, whose limbs are still quivering in the throes of death.[8]

What Fabre observed and at first interpreted as a murder turned out to be a sneaky kidnapping. The digger wasp inserted her stinger into her victim three times, injuring the three nerve centers responsible for movement so severely that although her victim was not dead, it was completely paralyzed. The cricket had become a package of live food that would stay fresh until the larvae in the wasp's burrow hatched from their eggs.

Fabre discovered this secret, too. He dissected a cricket and found the exact spots where the digger wasp had inserted her stinger. She had targeted what are known as the ganglia, bundles of nerves in the insect's nervous system. Fabre was dazzled by nature's know-how. How could it be that a small wasp driven by instinct "understood" more about the anatomy of a cricket than most scientists of his day?

I have said that the sting is driven several times into the patient's body: first under the neck, then behind the prothorax, next and lastly towards the top of the abdomen... But open a Cricket. What do we find to set the three pairs of legs in motion? We find what the Sphex knew long before the anatomists: three nervous centres at a great distance one from the other. Hence the magnificent logic of her needle-thrusts thrice repeated. Proud science, bend the knee![9]

A brilliant observer and writer, Jean-Henri Fabre went on to discover much more about predatory wasps. He watched as other species of wasps attempted to get the digger wasp to release her prize. He was astonished to discover that all parts of the digger wasp's complicated maternal behavior were driven by instinct alone: from excavating her burrow to tracking down prey and the surgical precision with which she immobilized it. Fabre was the very first scientist to show how complex and extraordinary the behavior of such supposedly simple insects could be.

It's a good thing I've read my Fabre, I thought. After that, thanks to the internet, it was all child's play. Within the huge family of digger wasps, I found what I was looking for.

Sceliphron curvatum, an Asian mud dauber wasp, arrived in Europe from India and Nepal and has been spreading since the 1990s. This new arrival specializes in snatching tiny crab spiders that hang out in colorful flowers. It captures and paralyzes these arachnids so it can use them as a food source for its young. Unlike Fabre's yellow-winged digger wasps, mud daubers do not dig brood burrows. As their common name suggests, they gather particles of clay in muddy puddles. They then mix these particles with their saliva and form artistically shaped pot-like structures about 0.75 inches to 1 inch (2 to 3 centimeters) in size. They stuff a large quantity of zombie spiders inside and lay their eggs on top. You can probably guess what happens next. Immediately after they hatch, the mud dauber larvae begin to eat their way through their supply of spiders until the larvae finally metamorphose into a new generation of winged spider hunters.

~~~~~~

I HAVE TO admit, I was very proud of myself. Perhaps my biology studies were somewhat useful after all. I was sure I was going to impress my friend with my knowledge. Would she recognize my potential as a great biological detective? I was convinced she would thank me for the good news I was going to give her. I made myself a cup of coffee to celebrate the moment and dialed Susanne's number. I didn't even wait for her to say hi.

"I've solved the mystery," I blurted. "You have mud dauber wasps. No need to worry—they're not flour moths or anything nasty like that."

"Um..."

"And the biology of these little insects is absolutely fascinating," I continued excitedly.

She attempted to get a word in. "Frank—"

"And way back when, the French naturalist Jean-Henri Fabre—"

Susanne was gradually losing patience with me. I was sensing she did not share my enthusiasm for mud daubers and French naturalists.

"Frank! I've just cleaned up the whole mess."

7

The Cockroach Tsunami

The search for the miracle extract continued. What was it that made its own excrement so irresistible to *Blattella germanica*, the German cockroach? We were gradually narrowing down which ingredients attracted the cockroach nymphs to the test strips in the aggregation experiment and which did not.

The Bayer chemists were especially intrigued by a group of organic substances they had detected in *Blattella* excrement: carboxylic acids with exotic-sounding names such as myristic acid, enanthic acid, and elaidic acid. A few gave my colleagues at Bayer hope, as my experimental baby cockroaches appeared to be quite interested in them.

The path of progress, however, was not without its challenges. No matter how careful we were, and despite the three rows of electrified wire around the upper rim of the cockroach incubator, a few insects always managed to scamper out from their designated living quarters. Whenever one of our cockroaches ran through a neighboring lab, complaints came pouring in.

"Keep a closer watch on your charges! Or, better yet, just give up on your cockroach research, before your test subjects leave their deposits in our petri dishes!"

Our experiments did not endear us to our neighbors, and it was not easy to mollify our critics. The situation was worst in the climate-controlled cockroach breeding room, where it was always comfortably cozy. Unfortunately, we now knew the red incubation container was not the high-security prison we had hoped it would be. It was clearly no cockroach Alcatraz. Time and time again, *Blattella* nymphs and adults with and without egg packets overcame the electrified defenses and the watery moat. Then they settled in hidden corners, behind storage shelves, and between bins and boxes. This was not acceptable. No matter how fascinating I found these little creatures, I knew we had to kill any we caught in the confines of the breeding room.

My professor, Martin Dambach, and I tried to find a way to control the cockroach plague. Then we had an idea. What if we sealed the door of the breeding room and turned up the heat until the cockroaches died of heat exhaustion? It might not have been the best plan in the world, but we had to do something.

We got our hands on two powerful heaters and rolled them into the room. There was a glass peephole in the door.

"We need a control so we don't open the door too soon," Dambach decided. He got some sticky tape and attached a small transparent plastic box and a thermometer to the inside of the peephole. He put five or six cockroaches into the box and snapped the lid shut. Our control was ready. Then we turned the heaters on full blast, closed the door, and waited.

After an hour, we checked to see how many of the control cockroaches were still alive. The heat had done its devilish work. Nothing was moving in the little plastic box. We decided to venture into the room to check things out. Dambach pushed down the handle and opened the door outward. No sooner had he cracked it open than a brown mass rushed toward us.

Hundreds of cockroaches were fleeing their sauna-like prison. A wave of chitin flowed over the floor and up the walls. A short way down the corridor, the cockroach carpet started breaking up into smaller and smaller groups. What had started as a homogenous mass became hundreds of individual six-legged insects, first in the corridor and then in the stairwell, escaping to all floors and into all the laboratories and offices of the Institute of Zoology at the venerable University of Cologne.

The first thing I thought of was Roman Polanski's film *The Fearless Vampire Killers* (also known as *Dance of the Vampires*). His two goofy main characters are fighting evil but, in the end, their bumbling ensures that it spreads throughout the entire world. In my case, my professor and I had unleashed a tsunami of cockroaches that would likely flow into every room in the institute.

~~~~~~

WHAT HAD HAPPENED? How could the cockroaches have survived the inferno? Clearly, the door to the breeding room did not seal as completely as we had thought. A wisp of cool air snaked its way inside through a tiny crack. The cockroaches, scared by the deadly heat, must have left their hiding places, deployed their antennae, and made for the lifesaving draft. "Taxis" is the name animal behaviorists give to goal-oriented movement in the direction of an attractant. In our case, thermo-taxis led to the mass migration of insects to the crack in the door of the breeding room. There, in the breath of cool air, the cockroaches survived, while their poor fellow cockroaches in the control box did not.

There it was again, *Blattella germanica*, the super-pest, the perfect bug, equipped with exquisitely attuned senses and life-saving reflexes, and flat enough to fit into any crack that might offer it safe harbor.

Had the word *shitstorm* been in use in Germany at the time (it didn't reach there until much later), we undoubtedly would have unleashed one. Over the next few days, we had to calm many colleagues, from postgraduate students to the head of the Institute of Zoology. I was within a cockroach hair's breadth of not being able to complete my thesis. Luckily, the deadline was fast approaching. My most important result was that a mixture of four carboxylic acids—myristic acid, enanthic acid, phenyllactic acid, and dodecanedioic acid—was almost as irresistible to the baby cockroaches as the extract made from their feces. There was a high probability of them aggregating on a test strip sprinkled with this mixture. Individually, each acid was less successful in attracting the cockroaches. Bayer applied for a patent for the miracle mix under the title "Agents for Controlling Cockroaches." Unfortunately, I discovered that, as a student, I had no rights to my discovery. I could not, therefore, hope for any windfall from the sale of the irresistible cocktail. Student life is a constant struggle, alas, and my cockroach rental business looked as if it was going to be the only one of my endeavors to show any returns.

My sideline in leasing cockroaches, however, was also coming to an end. My friend had suddenly had enough of the dynamic young music station VIVA and had handed in his resignation. After that, there were no more lucrative contracts for cockroaches. Until, one day, I spotted a card on the noticeboard at the Institute of Zoology. It read:

Hello: I'm an editor at VIVA and I'm looking for the student who supplied my predecessor with cockroaches for films.

I'm continuing this project. However, I will NEVER touch these disgusting insects. YOU MUST be there at all future filmings!!!

I called the new editor immediately. He seemed confused. "What? You're the guy? Someone else just called and offered his help." A fellow student, suspecting a lucrative business, had gotten to him ahead of me. But with me or without me, *The Roaches* soon went off the air, and any hope for income from that source finally dried up.

As it turned out, neither my professor nor Bayer was going to get rich on my discoveries. The commercial exploitation of the carboxylic acid mixture proved to be somewhat challenging. The attractants lost their allure when Bayer's product designers mixed them with pesticides. To get around that, they designed a cockroach trap with two chambers. One contained the attractant; the other contained the pesticide. That preserved the efficacy of the carboxylic acid cocktail, but the selling price of this more elaborate trap shot sky-high. Bayer's marketing department reported back that the increased efficacy of the product did not justify its increased price. The carboxylic acid cockroach trap never made it to the marketplace.

8

What Are Cockroaches Good For?

My cockroach year was coming to an end. I had written up my thesis and handed it in. Now I needed to celebrate.

"What are cockroaches good for?" a friend asked at my modest thesis party. At first, I didn't understand the question.

"What do you mean, 'good for'?"

"Well, do they have a purpose?" She narrowed it down for me. "Are they useful for something the way bees or earthworms are?"

An interesting question, and one I had never once asked in all the months I'd been working with them. Do animals need to be useful, sweet, clever, or dangerous for scientists to be interested in them? Biology means the science of living things. Biologists want to find out how life arose; how the diversity of species evolved; and how animals, plants, and other life-forms grow, reproduce, and, in the end, die. What's most fascinating is which aptitudes and survival strategies enable an organism to be successful and which do not.

To answer my friend's question, let's look at things from a different perspective. For a moment, imagine you are a cockroach.

A cockroach very likely has no idea it's a cockroach. It doesn't wonder about its role in the universe or if it's important that its species survives. Those are questions humans ask. Biologically speaking, a cockroach is the sum of all features that make a cockroach. These features include, for example, a flattened body, six legs, aggregation behavior, compound eyes, egg packets, speed, and so on. All these features are written into the cockroach's genes and passed down from generation to generation via eggs and sperm. The genes are immortal as long as a cockroach and its offspring successfully reproduce. The body that houses these genes dies and decays when an individual cockroach dies. But if the genes are immortal, why do they need all these physical trappings, whether in the form of a cockroach, a sperm whale, or a giant redwood?

The answer is that without a body to live in, genes would die. Chemically speaking, genes are enormous molecules of deoxyribonucleic acid—or, DNA. Absent a body to protect them, enormous molecules are not stable. They would be degraded by solar radiation and chemical processes. Within cells, however, DNA is constantly repaired, copied, and increased by an army of enzymes. These processes demand energy.

This is where the protective structure—in this case, the cockroach's body—comes into play. Its main purpose, from the genes' point of view, is to constantly gather food to fuel the powerful apparatus that ensures DNA not only persists but also proliferates and gets passed down to the next generation. To make sure this happens, the cockroach body finds a sexual partner in order to create offspring where the DNA can live—and so the genes survive. When a cockroach eats a bread crumb that has fallen on the kitchen floor, there is no deep meaning to this action. The physical expression of the cockroach is simply acting out a program written in its genes that has a single goal: keeping the genes alive. When a cockroach leaves its hiding place at night to find something it can eat, it's not foraging to ensure the survival of its species—*Blattella germanica*, for instance—but to make its genes immortal.

Interestingly, the parental care exhibited by some cockroach species fits into this concept of the "selfish gene"[10]—even though at first glance this behavior appears to be selfless. In the end, when a female cockroach feeds and protects her nymphs, she is also feeding and protecting her genes, genes that her nymphs have inherited.

~~~~~~

DESPITE THIS PROGRAMMING, some species die out while others thrive. There are reasons why this happens. For all life-forms, including cockroaches, the world consists of resources. Life-forms compete for these resources, with others of their own kind and with other species. Trees compete with other trees for sunlight. Cockroaches compete with other cockroaches and with other insects, birds, and mammals for food and places to hide. Lizards hunt cockroaches, and parasites attack their hosts. In the end, the winners in this powerful competition are the organisms (and therefore the genes) that exploit available resources most successfully.

Those that succeed are not necessarily the largest, the most dangerous, or the most intelligent life-forms—as cockroaches attest. Their blueprint is a winner, even without fangs, venom, or an enormous brain. They are omnivorous and their flattened body fits into the smallest hiding place. The senses of these little insects are exquisitely attuned to the slightest vibration or disturbance in the air when a predator approaches. And, as mentioned, some even care for their young.

The blueprint for "cockroach" has remained almost unchanged since the Paleozoic era—that is to say, for the past 300 million or so years. Over the course of these 300 million years, there have been two far-reaching extinction events, in which the majority of species living at the time disappeared from the face of the Earth. The cockroach, however, has prevailed, while the footfalls of the mighty *Tyrannosaurus rex*, for instance, only reverberated around our planet for a couple of million years—and the mammoth, the saber-toothed tiger, and the flightless dodo all disappeared long ago.

~~~~~~~

IF YOU WANT to pursue the "point" of cockroach existence further, you could argue that many cockroach species fulfill a not-insignificant ecological function. As what are known as decomposers—in their case, shredders—they help convert dead life-forms into nutrients that sustain new life. They are also part of an enormous food web in which they are prey for other animals that give us a great deal more pleasure: birds, reptiles, and insect-eating mammals. And so, cockroaches are indeed "good" for something.

It is, however, highly unlikely that any species other than humans pencil out this kind of cost-benefit analysis. I propose framing the question very differently. Not "What are cockroaches good for?" but "Why does the totality of genes that are a cockroach perform so well in the grand struggle for resources?" For those who approach the question this way, there is no better or worse, no useful or harmful life.

~~~~~~~

DOES THAT MEAN that there is also no higher purpose in life for humans than to pass down our genes? From the point of view of an evolutionary biologist, does it matter whether humans are a successful model, like the cockroach, or an evolutionary dead end that will disappear from Earth as quickly as we made our triumphant appearance and blink out after just a few hundred thousand years?

I find that kind of thinking shortsighted. In the end, humans are more than simply a "naked ape," as the British zoologist Desmond Morris called *Homo sapiens*. What makes the difference is our powerful brain. With its help, we can express complex concepts in words and gather knowledge, write it down, and distribute it. Human existence encompasses politics, arts, love, war, and religion, to name just a few of our cultural "achievements." Thanks to our high-performance brain, our behavior is extremely complex and adaptable, allowing humans to conquer just about every corner of this planet (with *Blattella germanica* and many other housemates in tow). I can, therefore, understand those who consider *Homo sapiens* to be by far the most interesting species on this planet.

Analyzing biological systems can't explain everything. Think of an opera by Mozart, or technological feats such as the invention of the wheel or the computer. We will never find a gene that can write a Monty Python sketch, or a rap song, or a new basic law of the universe. Through culture, humans have become more than simply packages of preprogrammed genes. The irony of our story is that our brain has made us a species that invented genetic engineering methods with which we can intervene in the course of evolution and "optimize" living beings according to our own needs. We are as much in charge of our own genes as they are in charge of us. Humans really are a crazy species.

BUT WHAT DOES all this have to do with the question "What are cockroaches good for?" I admit, I got a bit off topic. However, despite how special we are with our culture and our science, sometimes it's worth climbing down from our high horse and not using ourselves as a yardstick to measure every life-form or try to find a higher purpose for everything. Life finds a way, an outer form, and a survival strategy. Whether it's a chimpanzee or a cockroach, the ultimate test is which model is more successful. Unfortunately, I have to say that, thanks to human activity, the prospects for chimpanzees are considerably darker than those for *Blattella germanica*.

PART II

Return to the Rainforest

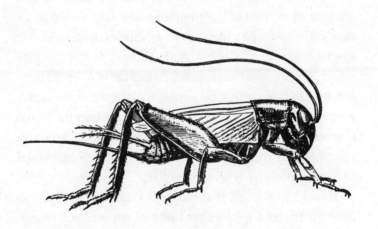

9

Answering the Call of Love

n 1912, the Slovenian behavioral scientist Ivan Regen
undertook an unusual experiment. The biologist was
researching the chirps of male field crickets. He suspected
that the *crii, crii* of the singing insects served to attract a
female ready to mate. This assumption made sense. Male field
crickets sit in front of their burrows and sing almost inces-
santly. They rub their wing covers against each other so that
special membranes in their wings begin to vibrate and radiate
sound. Sometimes a female comes by and the two insects mate.
But does that prove that the female is attracted by the male's
song alone? Could scent also play a role? Or perhaps the female
is reacting to the male's appearance? The song might have a
completely different function. Maybe it drives off other males
that might potentially compete with the singer.

It was highly likely that the male cricket was singing in
order to attract a female, but this had not yet been proven.
It was just a theory. A suitable experiment was required for
reliable proof. This is how science works, and Regen was com-
mitted to this process. To prove his theory, the biologist had
to exclude all other possible stimuli emanating from the male

cricket that might be the source of the female's interest. His female test crickets must not be allowed to see or smell the singing male, or be able to detect him in any other way. In order to prove his hypothesis, Regen wanted to present the insects with the male's song alone.

He had a problem. Back in 1912, when Regen was doing his research, there were no cassette recorders or any other kind of specialized equipment he could use to record the cricket's song and then play it back to a female. What did exist, however, was the telephone. Regen decided he would transfer the chirp of a male cricket from one laboratory to another via the phone line—a live call from a male cricket to a female cricket. What sounds simple to us today was a complicated and fiddly operation for the entomologist. He explains the challenges he faced in the paper he published documenting the experiment:

> There were problems at the outset because a regular telephone does not transmit the stridulation [i.e., the chirping] loudly enough and therefore I first had to find equipment suited to the purpose.[11]

Regen experimented with homemade loudspeakers and tin-can telephones. Finally, his innovation and perseverance were rewarded. When the chirping of a male cricket rang out of his telephone handset, the test female began to move and ran purposefully to the source of the song. What had begun as an assumption had been affirmed as a fact: cricket song triggers a goal-directed movement on the part of the female toward the source of the sound—what is known as phonotaxis (like the thermotaxis that helped the cockroaches escape the inferno in the breeding room). Incidentally, the song of the male cricket is known as a "calling song," but the fact that you make a telephone *call* is just coincidence, and the name "calling song" has nothing to do with Regen or his ingenious experiment.

REGEN'S TELEPHONE EXPERIMENT heralded a new direction in scientific research. The goal of this research was, and still is, to decode the meaning and messages in the sounds animals make. This branch of biology is called bioacoustics. My professor, Martin Dambach, was a bioacoustic researcher. The cockroach research was just one chapter in his scientific life story. He, like Regen almost one hundred years before him, was considerably more interested in the acoustic communications of insects, especially crickets. He asked me if I would like to stay in his laboratory for my doctorate. If I did, I would need to switch my interest to crickets and bioacoustics.

"I'd love to keep working with you," I answered in all sincerity, "but as you know, I'd really like to go to the tropics for my doctoral work."

Without hesitating, he told me he thought something could probably be arranged.

~~~~~~~

PROFESSOR DAMBACH KEPT his word. All it took was a couple of telephone calls and a few months later I was traveling to South America for a second time. Unlike my first trip, when I embarked on the ill-fated hummingbird project, my destination this time was not Colombia but the neighboring country of Ecuador. From the capital city of Quito, high up in the Andes, I boarded a domestic flight that was to take me to the research area.

From the window of the airplane, I saw the snow-covered peaks of the Andean volcanoes Cayambe and Antisana rearing up out of the clouds—these peaks are over 18,000 feet (5,500 meters) high. The mountains then quickly diminished in size, until an enormous plain stretched under the airplane. I was finally arriving in the "Wild East" of Ecuador—the *Oriente*, as the Ecuadorians call this part of the gigantic Amazon Basin.

The jet landed at the airport of the small town of Lago Agrio. As soon as I stepped out, the warm and humid midday heat set my sweat glands to work. I was back in the tropics. A short while later, everything seemed familiar as I found myself once again in a cheap hotel room, just as I had been two years earlier in Colombia.

10

~~~

Welcome to the Rainforest

What is wrong when everyone, including its own inhabitants, calls a town officially named Nueva Loja "Lago Agrio," which means "sour lake"? The little town in question lies in northeastern Ecuador, in the most northwesterly section of the Amazon Basin, in the largest contiguous rainforest in the world. In 1972, the oil company Texaco began to extract oil from around Lago Agrio. A small town grew up in the rainforest province of Sucumbíos. The oil companies came first, and with them, the first simple roads. Then, settlers followed from all over Ecuador. Slowly and then with increasing speed, oil fields, villages, and plantations began to eat their way into what had, up until then, been mostly undisturbed rainforest.

In 1995, I slept in Lago Agrio for the very first time. The Grand Hotel in the little town was, despite its promising name, a plain concrete structure—although, at three stories, it was one of the tallest buildings in the area. My room was sparsely furnished with a bed, a chair, and a lamp. That was it. I was to spend my first night in Amazonia, the place of my dreams, within the confines of these four bare walls. Footsteps thudded down the hotel corridor and a penetrating odor crept into my room—a hint of crude oil was infiltrating every corner of the hotel. Those must be workers coming home from the late shift, I thought as I lay on my bed. I closed my eyes and felt the sweltering tropical air. Lago Agrio, with its oil depots, fuel tankers, and workers, reminded me of Wesseling, the small industrial town on the Rhine where I had grown up. I reminded myself that this oil town was only the jumping-off point, the place from which I was to start my rainforest adventure.

My professor had delivered on his promise and introduced me to one of his colleagues. Klaus Riede, sleeping in the room next door, was a bioacoustic researcher like Martin Dambach, but Riede specialized in the tropics. When he was not observing the behavior of grasshoppers, he was conducting his research armed with a microphone and tape recorder. Before coming to Ecuador, he had recorded sounds in the rainforests of Malaysia. Back then, it was a brand-new idea to define entire ecosystems by their sound, and Riede was a pioneer in the field.

Riede used computers and specialized software to analyze his recordings and estimate the abundance of species in his research areas. Many different animal sounds on the recording suggested that a habitat was not only full of animals that communicated acoustically, but probably equally rich in life-forms that kept quiet. The acoustic profile of an area was supposedly an indicator biologists could use to estimate the area's overall biodiversity. The idea was that, armed with a microphone and a computer, biologists could identify biodiversity hot spots: places with a particular wealth of species that were natural paradises worthy of protection. And not only that. If researchers repeated their recordings at different times—during the day, at night, in winter, in summer, and during rainy or dry periods—they could discover a lot more about an ecosystem. The sound of a region would reveal whether the lives of its inhabitants followed diurnal, nocturnal, or seasonal cycles.

At first glance, this question does not seem particularly interesting for a woodland in temperate climes. Few animals are active in the winter, but in spring everything wakes up and most species begin their acoustic advertisements for mates. In a woodland like this, life explodes in spring. All you have to do is listen to understand that. Our knowledge of tropical rainforests, however, was at that time very patchy. Riede's sound recordings were going to help answer many questions.

More than twenty years later, the methods used in bioacoustic research have become much more sophisticated. That is especially true when it comes to analyzing digital sound recordings. These days, computers running specialized programs can identify which animal sounds are being played back to them. These programs work in the same way as music recognition software like Shazam. If you want to know what song is being played on the radio, you whip out your phone and start the app. The program compares the unknown piece of music with an enormous music database that it accesses online. When Shazam finds a match, the app tells the user which piece, by which artist, they are now listening to. Similar programs have been developed for animal sounds. The app Warblr, for instance, advertises that it can, like a sort of ornithological Shazam, identify more than eighty species of birds that live in the British Isles by their song, with a success rate of over 95 percent.

~~~~~~

"SOUNDSCAPE" IS THE name biologists give to a sound recording across a habitat, a hybrid word that replaces "land" with "sound" in the word "landscape." The dream of some conservationists is to increasingly automate the process of recording and interpreting soundscapes. For instance, you could install microphones in a conservation area that send their recordings directly to a computer. The computer could then continually analyze and compare soundscapes. A system like this would alert conservationists and biologists when something out there in the wild changed. Even if today's technology cannot yet identify all animal sounds with certainty (especially within a complex mix of animal sounds), the system could still raise the alarm if, overall, fewer animals were singing, individual songs were disappearing, or new ones were being registered. The more reliable and detailed these automated soundscape analyses become, the sooner conservationists will have an invaluable tool to monitor a particular habitat.

We have come a long way since those early days. A team of biologists in Costa Rica undertook an exhaustive field study that demonstrated how accurately automated sound analyses can capture the biodiversity of a tropical forest.[12] In 2015, ornithologists Mónica I. Retamosa Izaguirre, Oscar Ramírez-Alán, and Jorge De la O Castro installed recording devices in twelve locations in Santa Rosa National Park in Costa Rica. The researchers not only analyzed the soundscapes from their chosen spots but also recorded bird species using traditional methods. Established estimates of bird numbers within Santa Rosa National Park indicated that some zones had more bird species than others. They also showed, unsurprisingly, that most bird species sing early in the morning. For the most part, the automated soundscape analyses delivered an identical result. And in the experimental plots with a higher diversity of birds, the recorded sound was more complex than in the plots with fewer species. The ornithologists' software also identified more birdsong in the morning than at other times of the day and night, which coincided exactly with the times they had observed the birds being more and less active. The results of the study showed how powerful automated sound analyses can be even today, and that with advances in digital sound recognition software, they could relieve biologists of much more fieldwork in the future.

~~~~~~

BACK IN 1995, however, soundscape research was still in its infancy. I was to assist Klaus Riede with his work and, at the same time, carry out my own thesis research. Now that I was here in the tropical rainforests of Ecuador, my plan was to continue what Ivan Regen had begun back in 1912 with field crickets in Germany. I wanted to find out how many different cricket species lived in a defined area of this rainforest, when and how they sang, and how they managed to attract a partner with their calling songs, surrounded as they were by dense rainforest and hundreds of other animal sounds.

Not everyone grasps why a biologist would travel halfway across the world to track down cricket songs, especially when the trip involves spending weeks on end in the rainforest without access to familiar comforts like a soft bed. I, however, was thrilled by the project from the get-go. A biological treasure slumbers unnoticed in tropical rainforests, and I absolutely wanted to help rescue it from obscurity. Crickets are similar to cockroaches and almost all other orders of insects: most species do not live in Europe or the temperate regions of North America, but in the hot, humid tropics. For every European species, hundreds, if not thousands, of species are living out their lives in the tropical jungles of Asia, Africa, and Central and South America.

No one knows exactly how many species of animals and plants inhabit Earth today. U.S. entomologist Terry Erwin was the first person who tried to answer this question in a systematic way back in the early 1980s.[13] His methods were, it has to be said, somewhat drastic. The young tropical researcher was working in the rainforests in Panama. One obstacle constantly stood in his way: most insects in tropical forests live in the canopy. But research in the treetops—nearly one hundred feet (over thirty meters) above the forest floor—is difficult, expensive, and dangerous. You have to climb the primeval giants while keeping an eye out for snakes, wasps, and ants, and, of course, taking care you don't fall from a great height.

Erwin therefore decided to spread huge nets out under the crowns of tropical trees. He restricted his study to trees of a single species, an evergreen in the Malvaceae family called *Luehea seemannii*. Erwin's goal was to catch as many of the insects that were living in the *Luehea* crowns as he could. The only way he could do this, however, was by using a particularly brutal method. Erwin opted for chemicals. He sprayed the *Luehea* trees—then he waited to see what insects rained down into his nets. He called this "fogging."

Although *Luehea* trees are not that tall and only a few other plants such as bromeliads, lianas, and orchids were growing on the branches of the trees he selected, his results were mind-blowing. On only nineteen *Luehea* trees, Erwin found over one thousand species of beetles. Starting with this incredible diversity, Erwin began to crunch the numbers. The young biologist estimated that about 160 of the beetle species were specialists that lived exclusively in *Luehea seemannii* trees. Botanists, however, know of about fifty thousand species of tropical trees. Because Erwin also assumed that about 40 percent of all arthropods—insects and their direct relatives—are beetles, he estimated on the high side, and at the end of his calculations, he came up with an absolutely unbelievable number. Erwin's experiment suggested that tropical rainforests harbor about 30 million species of arthropods (insects, spiders, centipedes, millipedes, and terrestrial crustaceans). So far, biologists have discovered and described about one million insect species. Erwin's calculation suggests that about 29 million more species of arthropods are waiting to be discovered by science.

Erwin's work stunned tropical ecologists and was intensely debated. Were Erwin's estimates perhaps too high? Were there "only" two, five, or ten million undiscovered species of insects? No one has yet been able to answer that question with any degree of certainty. All we know is that species diversity in the tropics is truly immense. Especially when you consider that each insect is habitat for all kinds of other life-forms: gut bacteria, parasitic protozoa, mites, fungi, and many more.

What is it that makes tropical forests so biodiverse in the first place? This question, triggered by Erwin's research, suddenly became the most important question on the minds of many tropical biologists. And I, too, wanted to help answer it. I was convinced that a wealth of new, as-yet-undescribed species of crickets awaited me in Ecuador—and with every species there would be a new cricket song. My hope was to come across species with particularly special songs and adaptations, species whose existence cricket researchers had never even dreamed of.

~~~~~

IT IS ASTOUNDING that so many animal species share rainforest habitats. They must somehow divide up the resources they are competing for in the great struggle for survival. Singing crickets have a special problem: when lots of males of different species are in the same place trying to attract females with their song, there must be some kind of mechanism to ensure that the whole thing doesn't devolve into chaos. The insects need to make sure that a female from one species—attracted by the song of another—doesn't crawl to the wrong male. Or to put it in more scientific terms: the different cricket species must divide up the available bandwidth, the soundscape, among themselves. I wanted to find out how they do it.

Of course, these questions were not even remotely what was keeping me awake in my spartan hotel room. What was worrying me was the thought that I might not be able to achieve what I had set out to do. What if I failed in my work in this foreign country, in an extreme climate, surrounded, as I imagined it, by clouds of mosquitoes?

The next morning, we were going to depart for a small village without any roads leading to it, that was not connected to any electric grid, and where cell phones didn't work. I was headed for a place where I would not be able to call a friend and arrange to go out for a beer should I be overwhelmed by homesickness. There would, however, be running water in the rainforest village of San Pablo de Kantesiya, my home for the coming weeks—in the form of the mighty *Río Aguarico*, which flowed into the even mightier *Río Napo* and from there eventually joined the *Río Amazonas*, the most powerful river in the world.

11
~~

The Village in the Rainforest

The next morning, we hired one of the many pickup trucks waiting for customers on the streets of Lago Agrio. Klaus Riede and I heaved our equipment, our luggage, and our provisions for many weeks into the truck bed. Then our truck taxi clattered and bumped its way out of town, along one of the few roads that, in those days, headed east out of Lago Agrio. As I watched the landscape streaming by, my mood changed from euphoria to disappointment to dismay. For long stretches, there was hardly a tree to be seen. The rainforest that must have once blanketed the landscape was gone. Instead, we passed a series of simple wooden houses, modest plantations, and cow pastures.

"This is *colonos* country," our driver explained.

Colonos is the name Ecuadorians give to settlers who come to the rainforest provinces from other parts of the country to clear new pastureland. This *colonización* follows the roads; in the rainforest province of Sucumbíos, the oil companies were the ones that cut the first swaths through the forest. From these main roads, a network of small side roads spread out into the once-undisturbed ocean of trees. Gradually, more and more settlers arrived in the Amazon and bit by bit the rainforest disappeared.

After an hour's drive through farmland, the bumpy road ended at a river, the Aguarico. We had not yet seen any sign of virgin forest. Instead, a simple hut stood on the riverbank. The hut turned out to house a small bar with a billiard table. Next to it stood a row of market stands, and a volleyball pitch where sweaty men were partying heartily. We had arrived at Poza Honda—"deep puddle"—the small trading post where our drive from town ended.

~~~~~~~

A FEW RIVER taxis had tied up on the bank of the Aguarico: canoes about thirty feet (about ten meters) long, hollowed out from single tree trunks and equipped with powerful outboard motors. Poza Honda, I realized, was at the intersection between Western civilization—with its roads, electricity, and houses—and a world inaccessible except by water along the huge rivers that flowed through the rainforest. In Poza Honda, people traded things that could be found in only one of these two worlds. Noodles, rice, beer, tools, pots and pans, and cosmetics—along with much more that was brought here from the highlands of Ecuador in rickety trucks—were handed over from behind the counters of the market stands. From their huge canoes, Indigenous Ecuadorians[14] offered freshly caught river fish and large chunks of tropical wood.

"Klaus, *cómo estás?*" someone called from one of the boats. This was not the first time my new boss had traveled in Ecuador's *Oriente*. The news was already out that Klaus and I were on our way to San Pablo de Kantesiya. Renaldo, the owner of one of the river taxis, was expecting us. He grinned broadly. The pickup was quickly unloaded, and our gear was stowed in Renaldo's boat. After a beer to welcome us, our new guide—about fifty, fit and muscled—fired up his outboard and steered his canoe into the river. The motor powered the dugout forward at an astonishing speed. On either side, murky river water sprayed and foamed. At the prow, a second man kept watch for branches floating downriver that might impede our progress or mangle the propeller on our outboard.

Barely twenty minutes into our trip up the Aguarico, the landscape changed abruptly. The cow pastures and small settlements that had edged the river for the first stretch now gave way to dense forest.

"We've just crossed into the territory belonging to the Siona-Secoya," Klaus explained. "They'll be our hosts for the next few weeks."

The change in the landscape was astounding. I was becoming more and more excited to see the rainforest village and meet its inhabitants. What were the Siona-Secoya, a small group of just a few hundred Indigenous people, doing differently from the *colonos*, the settlers from the Andean provinces? I would soon find out.

Finally, one hour and a few dozen river meanders later, a large clearing in the forest appeared. As we neared the bank, I could make out wooden huts built on stilts and, next to them, a schoolhouse, a soccer pitch, and even a small church. About fifty people of various ages were waiting for us on the bank. Klaus was greeted in a friendly manner. I, however, got a more critical reception. The whole village lent a hand and carried our things to an empty hut.

Our accommodation, like the rest of the huts, was raised up on stilts. The walls were made from halved bamboo stems, and the roof, constructed from artfully woven palm leaves, looked as though it could withstand the strongest tropical storms that might come our way. Klaus and I each had a spacious room. There was also a huge, airy veranda where I could hang my hammock. On an earlier trip, Klaus had used some of his research money to have our home built. A few men from the village had sawed boards, posts, and joists to order and erected the beautiful hut. However, because no one had been looking after the small dwelling in the months before Klaus's return, we had to be wary in case termites had nested in the building.

When it's a matter of defending their jungle homes from board-and-joist-eating arthropods, even entomologists occasionally make an exception and reach for the insecticide. We sprayed every wall and every corner of the hut with termite killer. That flushed out all the six- and eight-legged houseguests. Cockroaches of all sizes scurried out from every crack and cranny between the wooden floorboards and the bamboo walls. A tarantula that would have fit nicely in the palm of my hand crawled out of the woven palm-leaf roof. Even though my previous year studying cockroaches had toughened me up somewhat, I was sure of one thing: I was never going to be able to shut my eyes for a moment in this place.

But then I unrolled my sleeping mat and hung my mosquito net for the first time in the Amazonian rainforest village of San Pablo de Kantesiya. A simple mosquito net can work wonders in situations like this. Under the *mosquitero* I suddenly felt safe and hidden. I relaxed and listened to the chorus of animal sounds ringing out from the surrounding forest. I heard the croaking of frogs and a strange screeching. And then there was a dull *whoop whoop whoop*. Somewhere out there, an enormous Amazonian cane toad was trying to bewitch a mate.

What I liked most was that right from my bivouac under the mosquito net I could hear a wide variety of cricket songs. A bright *tick tick*. A muffled, drawn-out *crrrrrrrr* and a *cri cri* that sounded like the song of the European field cricket, the subject of Regen's telephone experiment. From the other direction, from San Pablo's soccer pitch, came a din that had absolutely nothing in common with the pleasant song of the cricket.

It was a continuous, energy-sapping noise that reminded me of the sound made by a smoke detector that springs into action to save the residents of an apartment from a fiery death. Luckily, the soccer pitch was quite some distance away. It would have been impossible to sleep right next to this noisy creature. What kind of animal was responsible for a noise like this? Was it perhaps another cricket?

I was sure of one thing: I was in the right place. I was a doctoral student in bioacoustics. If I returned to Germany from one of the most biodiverse landscapes on Earth—from this paradise of animal calls—without any useful results, then I would need to find another profession, pronto. I wanted to begin work the very next day. The first puzzle I set myself was this: What kind of insect could be responsible for the terrible racket coming from the soccer pitch?

12

Tormented by Crickets

Three cheers for technological progress! Whereas Ivan Regen had to resort to homemade microphones and telephones in his experiments to prove the phonotaxis of field crickets back in 1913, all I had to do was pop batteries and a cassette tape into my portable digital recorder and attach a directional microphone and headphones, and I was ready to start hunting crickets. I realized that a grown man who visited the soccer pitch equipped like an audio surveillance expert from the CIA presented a somewhat peculiar sight for the people who lived in San Pablo, but that is how we insect lovers roll.

I was reminded of what a student in my semester told me about a university trip to the South of France. One of the stops was Saintes-Maries-de-la-Mer in the Camargue, known for its bathing beaches. Even though the sun was burning brightly, the professor would not allow the students to swim in the warm waters of the Mediterranean. He instructed them to investigate the insect fauna of the beach and dunes instead.

"Frank, I have never been so ashamed," the student said when she was telling me about the excursion. "There we were, fully clothed and carrying butterfly nets and specimen tubes as we ran after beetles, flies, and ants and tried to avoid sunbathers—on a nudist beach!"

Luckily, my appearance in San Pablo de Kantesiya was not quite as bizarre.

The soccer pitch was a good place to practice before I ventured out into the rainforest. I pointed my microphone in the direction of the strange sound at the edge of the playing field and walked slowly toward the source. The continuous buzz got increasingly louder, which meant I had to keep lowering the recording level so the digital recorder didn't get overloaded. *Biiiiiiiiiiiiiiiiiiiiiiiiiiiiiiiiiiiiiiiiiiiiiiiiiiiiiiiiiiiiiiiiiiiiiiiiiiiiiiii iiiiiiiiiiiiiiiiiiiiiiiiiiiiiiiiiiiiiiiiiiiiiiiiiiiiiiiiiiiiiiiiiiiiiiiiiiiiiiiiiiiii iiiiiiiiiiiiiiiiiiiiiiiiiiiiiiiiiiiiiiiiiiiiiiiiiiiiiiiiiiiiiiiiiiiiiiiiiiiiiiiiiiiii iiiiiiiiiiiiiiiiiiiiiiiiiiiiiiiiiiiiiiiiiiiiiiiiiiiiiiiiiiiiiiiiiiiiiiiiiiiiiiiiiiiii iiiiiiiiiiiiiiiiiiiiiiiiiiiiiiiiiiiiiiiiiiiiiiiiiiiiiiiiiiiiiiiiiiiiiiiiiiiiiiiiiii.*

I was getting closer and closer to the strange singer.

~~~~~~~

ALLOW ME, FOR a moment, to take a leap in time and place to a small miracle and a historic thaw in political relations. In August 2015—during U.S. president Barack Obama's administration—there was a celebration to mark the reopening of the U.S. embassy in Havana, Cuba. After fifty frosty years, there was finally a détente between the two countries. U.S. diplomats moved from Washington to the Caribbean island country, where they were housed in apartments and hotel rooms. Everyone was hoping for a bright joint future for the U.S. and Cuba.

The détente was quickly followed by a severe setback at the end of 2016. A few U.S. diplomats in Havana were complaining of health problems. They were suffering from tinnitus, had problems concentrating, and were experiencing dizzy spells. The embassy staff reported high-pitched noises in the hotel rooms assigned to them that made it impossible to sleep.

It didn't take long for the U.S. government, now led by Donald Trump, to react. The U.S. administration suspected the Cubans were employing sinister weapons against the diplomats. Some people suggested the Cubans were practicing sonic warfare. Journalists from around the world asked experts whether an acoustic attack might explain the sudden health problems the embassy personnel were experiencing, what kinds of sonic weapons were on the market, and what effects these might have on potential targets. Might the socialists on the Caribbean island be using a long-range acoustic device (LRAD), a powerful loudspeaker that fires extremely loud noises directly at people like a cannon? Sound cannons are sometimes deployed on freighters traveling the high seas, for example, to keep pirates from capturing the ships.

But earsplitting sound cannons in the middle of Havana? The pulsing metropolis is certainly not the most peaceful spot on Earth, but surely someone would have quickly noticed a persistent, long-term attack on hotel guests. Was it perhaps a question of an acoustic bombardment in a frequency beyond the range of human hearing? Were infrasonic—that is, extremely low—sounds triggering the ailments? Other experts speculated that the U.S. personnel were being exposed to particularly high-frequency sounds—ultrasounds—or even to microwaves.

Astoundingly few experts, however, were asking why the Cubans wanted to harm their guests in the first place. What political advantage could they possibly gain from intentionally making U.S. citizens sick? But anything is possible when you're dealing with dastardly Caribbean socialists, it seems. The U.S. reacted and, citing security concerns, reduced its diplomatic presence in the country.

But then there was some progress in the international incident. One of the U.S. embassy's employees had recorded the noise that had robbed them of sleep. The Associated Press published a recording on their YouTube channel.[15] I am sure that anyone who has listened to this audio document can empathize with the suffering of the U.S. embassy staff.

~~~~~~

NOW LET'S ROLL the clock back twenty years. Back then, I was stalking my way across the soccer pitch in the small rainforest village of San Pablo de Kantesiya armed with a DAT recorder, a directional microphone, and headphones. The screeching I was tracking down was getting louder and louder. Then, suddenly, it stopped. That was a good sign. Most likely, I was on the trail of an insect, probably a cricket. A male that uses loud noises to advertise for a female has to expect that he will also attract other animals—insect-eating birds or mammals that would like to make a meal of him. Crickets are quite closely related to cockroaches and, like cockroaches, they have developed sensitive sensory organs that pick up the slightest vibrations from the ground and put the insects in a state of emergency preparedness. These mechanoreceptors, as they are called, are located in the insects' leg joints. The singing—or, I should say, *shrieking*—insect must have noticed me. I stood still, avoiding any further movement—and I waited. I don't know if any of the villagers watched as I stood motionless in the middle of the sports field bent over like a heron poised to catch a fish. That was of no concern to me. My hunting instinct was ratcheting up by the minute.

~~~~~~

THE SOUND RECORDING the Associated Press published on You-Tube was called up and listened to by a few people at first—then interest built, until thousands had heard it. Some commenters warned people against listening. "You will die," wrote one of

the all-too-curious internet users forebodingly. Another rec-
ommended the YouTube track to all who wanted to know how
awful it is to suffer from tinnitus. Yet another joked that the
sound they heard had to be the latest work of the British techno
artist Aphex Twin.

An international team of biologists also listened to the
recording on YouTube. These researchers, who worked with
amphibian specialist Alexander L. Stubbs at the University
of California, Berkeley, had been following the diplomatic cri-
sis. Reports by some of the victims of the acoustic attacks had
thrown up a red flag. According to the diplomats, the din had
mostly stopped, momentarily, when they moved about in their
apartment or hotel room, or opened a door. Details in their
accounts suggested to the biologists that insects, not sound
cannons, might be the source of the mind-numbing disturbance of the peace. They fed the YouTube recording into a
computer loaded with a software program that specialized in
acoustic analysis. The program created a sonogram, a graphic
representation of the recorded sounds.

It was immediately clear to the scientists that what sounded
like a continuous noise was actually a series of incredibly quick
pulses of sound, one after the other. To the human ear, these
individual pulses join to create a continuous sound, the way
individual frames in a movie are perceived by viewers as continuous motion. The researchers narrowed down the list of
possible suspects by analyzing the pitch of the sound and the
time lag between pulses. Eventually their analysis pointed to
a small insect about an inch (just a few centimeters) long: the
Indies short-tailed cricket, *Anurogryllus celerinictus*.

The YouTube recording from Havana was strikingly similar
to a recording of the calling song of this species tucked away in
the researchers' archive of insect calls. However, the bioacoustic
researchers wanted to be 100 percent certain before they

published their findings. After all, it's not often that experts in animal sounds get to play a role in an international political thriller. The acoustic analysis of the sounds from Havana certainly pointed to *Anurogryllus*, but there were also differences between the sounds from Cuba and the researchers' archival recording. The cricket in the researchers' recording, however, had been out in the wild and not inside a hotel room. Stubbs and his team decided to rerecord their recording of *Anurogryllus celerinictus* as they played it back in a closed room. Then they repeated their analysis using a recording of a recording. Now they were certain. What you could hear on the recording released by the Associated Press was *Anurogryllus celerinictus*—slightly altered by the echo as the call bounced off the walls of the hotel room.

Stubbs and his team presented their findings at a meeting for the Society for Integrative and Comparative Biology.[16] In their presentation, they underscored that they could only comment on the origin of the recording published on the internet. There might, of course, have been other attacks on the U.S. diplomats in Cuba. As far as I'm concerned, however, Stubbs and his team produced the most convincing theory to date to explain the sonic weapons affair in Cuba. The diplomats from the United States had been sickened not by a sound cannon pointed at them deliberately, but by a small—but loud—cricket.

～～～～

MY OWN STALKING of an insect's call on the sports field also ended in success. After I had remained motionless for a couple of minutes, the continuous sound started up once again, this time right next to me. By then, the sun had set and the small tropical village was cloaked in darkness. Agonizingly slowly, I bent down to get closer to the source of the sound. Then I turned on my flashlight. A short-tailed cricket was sitting in a small depression. The insects dig these depressions to focus the sounds they make and magnify them. The cricket that now cowered before me was not *Anurogryllus celerinictus*, the species that had unnerved the U.S. diplomats in Havana, but a close relative in the same genus. Its call, however, was equally as shrill as that of its Caribbean counterpart.

Something as simple as the calling song of a cricket can be all it takes, apparently, to unsettle newcomers to the tropics—especially if they are traveling from Washington or some other part of the United States with a slight suspicion of capricious Cuban socialists tucked into their baggage.

13

Opera in the Rainforest

Not every cricket species sings in tones as shrill as *Anurogryllus*. Quite the opposite, in fact. A friend who earns his living as a voice coach once told me, "The human ear WANTS to hear crickets. I always say to my students: 'Let me hear it; let me hear the cricket sing.'"

What he likely meant was that few animal sounds elicit such pleasant feelings in us as the calling song of a male cricket. The surest proof of this comes from the world of feature films. How often do sound editors play the chirps of crickets to signal that a love scene is on the way—even if the film is set in winter, or in places where you would never hear crickets sing? The chirps of crickets inevitably have us thinking of relaxing holidays in southern climes—and of romantic evenings with or without red wine. They are the dreamlike soundtracks of nature.

And yet the cricket is not that much different from the German cockroach, *Blattella germanica*—which never makes us feel romantic. Most cockroaches and crickets feed and behave in very similar ways. They eat all kinds of plant debris, live well-camouflaged lives in hidden spaces, and prefer to deal with predators by fleeing rather than attacking. Most people get agitated when they spot a cockroach, whereas crickets soothe us with their beautiful songs—with the exception of *Anurogryllus*, of course.

~~~~~~

AFTER I HAD solved the mystery of the noise on the soccer pitch, it was time for a real rainforest adventure. Klaus and I paid assistants from the village to guide us through the forest. After a few days, I knew where the beaten paths were that our neighbors used when they went out into the rainforest. I memorized the main intersections and ventured out on longer and longer excursions, which took me ever deeper into the tropical forest.

After a week, I felt safe enough to search out just the right spot for my first monitoring station. I found a small clearing created when a tree weakened by age fell in one of the many tropical thunderstorms that pass through this area. The thick trunk served both as a place to sit and as a table where I could set up my equipment. My plan was to spend twenty-four hours in this spot: a whole day and a whole night. I wanted to record a couple of minutes of animal chorus on the hour every hour so I could estimate which insects were active when.

The time had come for me to spend my first night deep in the rainforest. I was nervous because I wasn't sure how risky my plan was. I went through all the possible dangers in my mind. Would I be killed and eaten by a jaguar? No. These predatory cats are shy and also rare. Would I be bitten by spiders or stung by scorpions? This was also unlikely, especially if I wore stout shoes. What about venomous snakes? That was, unfortunately, something I did need to worry about.

The American lance-headed viper commands a great deal of respect from those who wander through the South American rainforest at night. These relatively large snakes leave their hiding spots when it gets dark, scouring the forest floor for small mammals. Unfortunately, the lance head is exceedingly well camouflaged, is highly venomous, and bites if you fail to notice it and get too close. On the other hand, venomous snakes are becoming increasingly rare in Ecuador because most people—our Siona-Secoya hosts included—kill the slinky predators when they slither across their path.

The greatest danger in the rainforest, however, comes surprisingly enough not from animals at all, but from plants. A tree here is not just a tree. Rather, it is habitat for hundreds more plants. Moss carpets sprout from its trunk, and vines twine their way up like ivy reaching for the sun. A succession of orchids, bromeliads, and other air plants, known as epiphytes, flourish and decay on its thick horizontal branches. Sixty-five feet (twenty meters) or more above the forest floor, some branches develop layers of humus so thick that seeds—blown in by the wind or carried in by birds, ants, and bats—are constantly taking root. The weight the branch carries increases. Eventually, all it takes is a heavy shower. When the small hanging garden is saturated with rain, even the strongest branch can no longer bear the weight. It breaks and crashes down to the ground. A projectile like this, of branch, humus, and epiphytes, can weigh four hundred pounds or so (a couple of hundred kilograms)—enough to swiftly bring an end to any doctoral research.

~~~~~~

ABOUT AN HOUR before sunset, I finished my setup in the small clearing and lay back to listen to the performance that, day after day and night after night, plays out in the great opera house that is the rainforest.

And with that, it's time to introduce the rainforest performers. The hours before dusk are when the cicadas take center stage. Cicadas are relatively large insects that cling to the trees' trunks and branches. They use their trunk-like mouthparts to suck sugary plant juices from the trees' vascular tissue. As with crickets, it is the male cicadas that attract females with their song. The insects use an organ along their sides called a tymbal to produce their characteristic humming sound. The tymbal consists of a membrane that is stretched taut, like the skin on a kettledrum. Muscles that extend from inside the body are attached to this membrane. These muscles contract and relax again incredibly rapidly. When they do this, the membrane starts to vibrate, and the characteristic hum of the cicadas fills the air. Thanks to the similarity of their tymbal to a percussion instrument, cicadas lay down the beat in this enormous tropical chorus. The males produce their sounds all day long, but they ramp up the action when the sun's rays shine through a gap in the clouds. As the sun sets, many species of cicadas synchronize their songs like a chorus belting out rousing renditions of the "Chorus of the Hebrew Slaves" in Verdi's *Nabucco* or the "Ode to Joy" in Beethoven's Ninth Symphony.

There are other percussive a cappella performers in the rainforest chorus. Woodpeckers sound like they are using chisels to hack holes in standing dead trees. They are looking for insect larvae eating the dead wood, and as they do so, they are also constructing their nesting cavities. The percussive tones also serve as a means of communication. Every species hammers out its own unique signal. Woodpeckers ram their beaks against extremely hard wood at rates of up to twenty times a second. In the mating season, that adds up to thousands of strikes a day. The birds drum with such strength that you can hear the blows from a couple of football fields away. The strong muscles around their beaks act like shock absorbers to cushion the impact, so the love-crazed woodpeckers don't fall senseless from the trees at the end of the day after having given themselves a concussion.

Other birds step up for the solos—the arias—in the rainforest opera. Songbirds produce their songs in much the same way human singers do. As they exhale, they set membranes in motion, just as Maria Callas once set her vocal cords in motion when she sang. In the case of the feathered singers, however, these membranes are located not in their throats but in what is known as the syrinx, close to their bronchial tubes. Early morning is a popular time for the birds to broadcast their solos, duets, and trios. Some species begin as soon as the first barely perceptible strip of light makes it over the horizon, then more and more feathered soloists join in. By the time the sun has risen, they are all twittering, trilling, and tweeting together like Marcellina, Figaro, Susanna, the count, and the countess in Mozart's opera *The Marriage of Figaro.*

~~~~~~

IN THE NIGHT, apart from the *whoo, whoo* of owls, you don't hear many birds. But that can change as soon as there's a full moon. My very first night in the rainforest was a bright night with a full moon. Punctually at sunset, the pale orb of Earth's moon rose out of the ocean of trees and climbed into the night sky. The moonlight helped me to relax a bit. It wasn't as pitch-dark as I had feared. Just as all my worries about falling branches and lance-headed vipers were fading away, a desperately sad and—at the same time—bone-chilling sigh rang out from the top of a tree directly over my head: *Wuhu huuuu huu hu hu hu hu!*

The cry sounded so human that at first the hairs on the back of my neck stood on end, and then I thought maybe one of the fun-loving villagers was playing a joke on me. The heartrending lament was repeated every couple of minutes.

Wuhu huuuu huu hu hu hu hu!

This was no breezy Mozart opera. It sounded more like Muddy Waters or B.B. King singing the blues. A few days later, I hummed the sound that had struck fear into every fiber of my being that night to our neighbor Renaldo.

"*Wuhu huuuu huu hu hu hu hu!*"

"Ahh. That's a bird. We Siona-Secoya call it *Paú*," he explained.

I had set up my monitoring station directly under a tree from which a great potoo was calling to the moon. Great potoos are about the length of your arm and in the same family as nightjars. They spend their days sitting completely motionless on a branch, so perfectly camouflaged that you can walk right past a potoo a hundred times without noticing it. Among ornithologists, the haunting laments of potoos have earned them the nickname "poor-me-ones."

<center>~~~~~~</center>

IN SOUTH AMERICA, many sad and spooky stories have grown up around this bird's nocturnal laments. Renaldo related one of them to me.

A long time ago, there lived a family—mother, father, daughter, and son—in a small hut in the middle of the forest. In those days, the moon didn't exist yet, so the nights were dark as pitch. All four were sleeping peacefully in their hammocks as they did every night. But suddenly the girl woke up. A man was pressing himself upon her and—oh well, you can guess what happens next...

(The Siona-Secoya are usually extremely reticent when it comes to the subject of sex.)

The visits continued night after night until it became too much for the girl. She wanted to know who was forcing

91

himself upon her every night. She spent the next day trampling the red fruits of a tree. She poured the crimson pulp into a bowl and set the bowl of red dye on the floor under 'her hammock. That night, the familiar drama played out. As her visitor once again lay with her, she quickly dipped her fingers into her bowl and ran them over the unknown man's face. The next day, the stripes on his cheeks revealed that it was the brother that sought the girl out every night. This incest had to be punished. The siblings were separated. The brother was turned into a bird and bound to a tree and the girl was banished far away up into the night sky—where she became the moon. Whenever her brother, now turned into a potoo, saw the full moon in the sky, he was overcome with guilt and melancholy and he couldn't help but cry out. And that is why the potoo's loud sighs sound so sad. *Wuhu huuuu huuu hu hu hu hu!*

OTHER ANIMALS SANG less tragic parts. Some mornings in the South American rainforest, you are woken from sleep by a loud barking. Groups of red howler monkeys are calling loudly to defend their territorial boundaries. "Don't come any closer. We're eating here!" is the message they are broadcasting to other howlers in the area. The monkeys' calls don't sound much like a gentle aria, but more like the battle cries of inebriated soccer fans—and they play a similar role, alerting rivals in an audible show of strength.

But back to the more soothing crooners in the rainforest opera. Frogs and toads make their entrance once darkness falls. Ecuador is a veritable amphibian paradise. Brazil and Colombia are the only countries where biologists have documented more species of frogs than in Ecuador, which is a considerably smaller country. Over six hundred species of frogs and salamanders hop and crawl through the rainforests in Ecuador. These include highly toxic poison dart frogs—which likely get their deadly toxin from what they eat—technicolor tree frogs, and giant toads weighing more than an adult rooster. The contributions made by the amphibians to the rainforest opera are equally varied, from the booming bass of the giant toad to the bright falsettos of the frogs, which are strikingly similar to the sound of a dripping tap.

And, of course, one mustn't forget the rainforest's experimental performers—bats. Human ears occasionally register some of their contributions as a very bright *click, click*, but most of the noises these winged mammals make are ultrasounds beyond our auditory range. Bats use ultrasound for orientation, and it works in much the same way as sonar. Numerous species of insect-eating bats use it to find their food: night-flying moths. But not all moths are defenseless against the aerial night hunters. Many moths have excellent hearing—even in the ultrasound range. If they are targeted by a bat, they drop to the ground to escape the attack. African cabbage tree emperor moths make themselves "invisible" to the bats' sonar in the same way a stealth bomber hides itself. The scales on the moths' wings create an extremely uneven surface—and they are even perforated in places. British scientists discovered that these scales are very effective at absorbing the bats' echolocation calls.[17] Tiger moths interfere with bat echolocation even more directly. They retaliate by sending out ultrasounds of their own, effectively jamming the bats' signals. The confused bats usually abandon their attempt to hunt down the annoyingly noisy moths.

This leaves us with the crickets. What parts do they represent? The answer is easy. They are the string section of the orchestra in this operatic production. These small insects produce sounds using the foremost of their two pairs of wings.

These leathery forewings are called tegmina. One of the wings scrapes the other so that gossamer-thin membranes on both tegmina start to vibrate—just as the strings of a violin are set in motion when the violin player touches them with their bow. The similarities between a string instrument and a singing male cricket don't stop there. One of the many veins on the underside of the tegmina is substantially thicker than the others. There are serrations on this vein, so tiny you can see them only under a microscope. This vein, known as a file, corresponds to the violin bow held by the musician. The cricket draws its file over the thickened edge of the other wing. The tiny serrations on the file get caught on and then released from the thickened edge, which sets the whole system in motion.

And that's not all. Like the strings on a violin, the wing membranes are "tuned" to a note unique to that species. They always vibrate in roughly the same frequency, which is known as the carrier frequency. The cricket's wings emit a very clear tone, consisting of the carrier frequency and its harmonic overtones.[18]

We find crickets so pleasant to listen to precisely because their wings are so exquisitely tuned. It does mean, however, that their setup for producing sound—their instrument, if you will—is far more restricted in range than a violin with its four strings. Indeed, the combination of membranes and file is so precisely tuned to the carrier frequency that the cricket can produce only a single note, but to make up for that, it can really amp up the volume. The advantage of a one-note system like this is that the cricket can produce astonishingly loud calling songs despite its tiny wings and muscles. And it is this ability that caused the aforementioned U.S. diplomats in Cuba so much distress.

In practical terms, the ears of the females are as specialized as the wings of the males. They are perfectly attuned to the carrier frequency of the males' songs, and so the females can clearly identify the song of a male of their own species.

If a female is to find a singing male, however, she has to do more than just identify his song. She must also be able to pinpoint the direction it's coming from, so she can track down her potential mate. Like humans, therefore, crickets have not one but two ears. However, unlike humans, their ears are not on opposite sides of their heads. Crickets hear with their legs. Their two ears, called tympanal organs, are located in the lower section of their forelegs. This exposed location might seem odd at first, but it makes perfect sense. It means the two ears are located far apart, which improves the female's ability to pinpoint where the sound is coming from. When the male is singing off to one side of the female, his song reaches her front leg on the side closest to him a fraction of a millisecond sooner than it reaches the leg on the other side. The two tympanal organs relay the sounds to the cricket's brain in the order in which they receive them. The brain relays signals to the cricket's legs. Like a mini-robot, the female now turns in the direction of the singing male and starts to move. (Unless the female cricket is the subject of one of Ivan Regen's experiments, in which case the goal of her phonotaxis would be a telephone receiver.)

But back to the singing male and his perfectly tuned tegmina. There is something that distinguishes these from the violin and its bow. Whereas the violin player can continuously slide the bow back and forth across the strings, after every pulse of sound, the male cricket must lift his file and put it back down so it engages once more with the thickened edge of the lower wing. Only then can he scrape the file over the edge of the lower wing once again, and only then can the next pulse of sound be produced. That all happens at such a high speed that the human ear is deceived. Each *criii* of the *criii, criii, criii* of a field cricket is made up of four to five single notes, which correspond to an equal number of wing strokes on the part of the male. The supposedly continuous tone of *Anurogryllus*, the cricket that so tormented the U.S. diplomats in Cuba, is in fact a series of hundreds, or even thousands, of pulses of sound.

The speed at which these pulses follow one another, known as the pulse rate, is relatively constant within a species, as are the characteristic pauses between a series of single pulses. Like the carrier frequency, the pulse rate is very similar for all singing male crickets in a single species, which suggests that these features are how females recognize the calling song of "their" mates.

This hypothesis was also tested. In the 1980s, some bioacoustic researchers refined Ivan Regen's experiment, this time using synthesizers to artificially create cricket songs. When these synthesized songs coincided with the carrier frequency and pulse rate of the natural song of the male, these manufactured calling songs triggered phonotaxis in the female test subjects.

This advance allowed the researchers to alter individual elements in the crickets' songs as they wished, which enabled them to unlock the secret of what exactly the females find irresistible. The results were unequivocal: female crickets respond to songs with a particular carrier frequency coupled with the signature pulse rate for their species—that is to say, the speed at which the singing males engage and disengage their wings, and thus how quickly the short pulses of sound follow each other.

~~~~~~

IT WAS GOOD to know what I should be listening for as I set out to understand how the many different species of crickets in the Ecuadorian rainforest avoided misunderstandings while communicating. On my first night in the rainforest, I was impressed by the incredible cricket string section playing all around me. On the forest floor, I heard a cricket with a low chirp, a double bass of the cricket world. There must have been a few cellos chirping in a slightly higher register on some of the tree trunks. A viola sounded out from the undergrowth. But the stars of the rainforest concert were the many delightful violins. A bright *criii, criii* rang out from all the trees, palms, and bushes. The forest must have been full of chirping insects, especially high up in the canopy. Most of these crickets "broadcast" at higher frequencies than I was familiar with from the field crickets back home. That very first night, I counted about twenty different cricket calls. Around me, hundreds of individual males were advertising for partners. Their calling songs melded together into a powerful symphony of nature. I realized there was going to be plenty of work for me to do in the forest around San Pablo de Kantesiya.

14

Red Legs, Green Legs

I s it really important whether a grasshopper somewhere in the Ecuadorian rainforest has red legs or green legs? This won't be a question of pressing interest to most people. Indeed, most people probably wouldn't waste even a moment thinking about it. For my boss Klaus Riede, however, the question was so exciting that he tasked me with organizing a small expedition. I was to take an hours-long boat trip up the Aguarico River to see what was going on. When he explained to me in more detail why this excursion was necessary, I wanted to leave immediately.

What scientific question lay behind this mission? To cut to the chase: like all biologists who study the tropics, Riede wanted to find out why such an unimaginable number of species live in tropical rainforests and how these species—most of them insects—came to be.

~~~~~~~

KLAUS IS AN expert on tropical grasshoppers. He is especially fond of one particular species: *Galidacris variabilis*, a pretty grass-green hopper with black markings. Unlike crickets—to which grasshoppers are related—and many other grasshoppers, *Galidacris* doesn't communicate using acoustic signals. Instead, to get themselves in the mood for mating, individuals wave to each other with their strikingly colored hind legs. Interestingly, *Galidacris variabilis* doesn't look the same across its range. Depending on where in Ecuador you are searching for these grasshoppers, you can find them with either red hind legs or green hind legs. In the northeast, you find only the red color variant, and in the southeast, only the green one. The Napo River, one of the many tributaries to the Amazon and Ecuador's widest river, appeared to be the dividing line between the two populations.

It is the different color variants that make these little grasshoppers so interesting. Is it possible that we are seeing evolution at work in *Galidacris variabilis*? Might a single species of grasshopper be in the process of splitting into two new ones? This was why Klaus wanted to know more about the little hopper. For example, we didn't know which color variant called the easternmost part of Ecuador home, or whether the Napo River was acting as a natural barrier between variants in that part of the country as well. To find out more, someone had to travel up the Aguarico River, deep into the mostly undisturbed rainforest of Ecuador. That someone was to be me.

Like Klaus, I was interested in the red-leg / green-leg problem, of course, but what was even more exciting to me was the prospect of traveling much farther into the Amazon Basin. In the village, I chartered a motorboat that came with Renaldo at the helm. I chose him because I had found him to be a prudent navigator on the rainforest river. Renaldo agreed immediately but told me we needed more help. And so, Javier joined us. Javier was one of the guides who had taught me how to keep from getting hopelessly lost in the Siona-Secoya forest.

THE THREE OF us squeezed into Renaldo's two-man boat at the break of dawn. This boat of Renaldo's was a sleek, high-speed metal bullet. The first thing our rainforest captain did was to steer us in what seemed to be the wrong direction. Instead of heading upstream toward the wilderness, we headed downstream toward the Poza Honda trading post. We landed there and the three of us climbed into the bed of one of the pickup-truck taxis waiting at the river's edge. The taxi drove us to the little oil town of Shushufindi so we could purchase food and fuel for the long boat ride. Renaldo filled a rusty old can with diesel fuel at one of the gas stations. While he was filling the can, he lit a cigarette. It was about 86 degrees Fahrenheit (30 degrees Celsius)—in the shade. Nervously, I asked the sinewy captain of my boat if it was a good idea to smoke while filling a can. Renaldo looked at me while he calmly continued to smoke. After a long pause, he said, "Relax. It's diesel. Diesel doesn't burn!" To prove his point, he held his lit cigarette even closer to the stream of fuel. My heart skipped a beat. Okay, I told myself, this might look like he's tempting fate, but perhaps he's just self-taught in the ways of petroleum products. After all, diesel does need to be heated up before it ignites.

~~~~~

A GOOD HOUR later, we were finally back on the wonderful Aguarico River, heading east. Unfortunately, after we left San Pablo and the territory of the Siona-Secoya, the forest on both sides of the river did not immediately get any denser or wilder. In fact, the trees began to thin out to accommodate large plantations and settlements. The *colonos* had penetrated much farther into the forest than I had realized. They managed their land very differently from most of the Indigenous peoples. The settlers clear-cut their fields, making swift work of the rainforest. The Siona-Secoya, Cofán, and Huaorani, in contrast, usually cleared only small sections of their territories to grow some sweet corn, manioc, or bananas. The *indígenas* cultivated their fields for just a few years before abandoning them to the forest once again. The Indigenous farmers then cleared a new small field of manioc, and nature reclaimed the small, abandoned plantation. The forest can heal in places like this. This style of farming, known as shifting agriculture, has worked for hundred of years without causing long-term damage to the forest. But the more people flock to the rainforest—bringing with them cattle and palm oil plantations—the more irreversible the destruction of the forest becomes.

~~~~~~~

NATURE IS, UNFORTUNATELY, not completely wild and undisturbed in the Indigenous territories either. However, even today, you can usually at least still find dense rainforest there—although progress exacts its tribute from Indigenous peoples too. For a long time, most *indígenas* hunted using blowpipes, spears, and bows and arrows. By the time I visited, however, in San Pablo de Kantesiya—as everywhere else in the *Oriente*—guns, sometimes even high-caliber hunting rifles, were all the rage. Everything that came into the sights of experienced hunters was in danger: monkeys, tapirs, deer, birds. It's much the same today. If someone needs money to visit the doctor or buy a new outboard motor, they go out and shoot a wild pig and sell their prize in Poza Honda and other similar trading posts.

My neighbor Renaldo and I were sitting on his veranda one evening, when he told me, with evident pride, that he had something he wanted to show me. He disappeared into the house and came back holding a pump-action shotgun, polished until it positively gleamed. All you have to do is move the slide forward and back again and you can reload a rifle like this in no time at all.

I was aghast. "What do you need a murderous weapon like that for?"

"When I found a pack of wild pigs in the forest, I used to be able to shoot just one of them. Now I can get three or four at a time."

People told me that once you could see whole herds of tapirs on the banks of the Aguarico River, lots of monkeys, and large flocks of birds shimmering in a multitude of colors. Those halcyon days, unfortunately, were long gone by the time I was clattering up the Aguarico with my two companions.

~~~~~~

AFTER AN HOUR of motoring through *colonos* land, the rainforest returned. We had finally arrived in the wilderness that I had dreamed of since I was a child. No settlers lived here. The only people you were likely to come across in this place were scattered families living in the rainforest, members of various Indigenous groups: Cofán, Siona-Secoya, and Quechua. Apart from these families, there was nothing but nature—a green ocean of trees that stretched all the way down to the river. The tops of the trees touched, forming a continuous canopy. Here and there, the occasional ancient giant 150 feet or more (more than 45 meters) tall rose up above the canopy. Flocks of long-tailed scarlet macaws—enormous red, blue, and yellow parrots—flew over our heads. The noise of our motor flushed raptors and colorful, long-billed toucans. It was the most wonderful boat trip of my life.

~~~~~~

AFTER TRAVELING FOR four hours, we tied up at the bank where a little hut stood, the home of a Quechua family. In the "Wild East" of Ecuador, people know and help each other. The welcoming family invited us to spend the night. We unrolled our sleeping mats on our hosts' veranda and hung our mosquito nets. When we were done, I was ready to go out to look for *Galidacris*, the grasshopper whose legs were sometimes red and sometimes green.

"Wait! You can't leave!"

It was clear I wasn't going anywhere. Anyone who lives five hours by boat from the nearest road and gets visitors only every couple of weeks wants to welcome their guests properly. The woman of the house brought a large bowl out onto the veranda, which we passed from mouth to mouth. The beverage we were drinking was *chicha*, an Indigenous "beer" brewed since the time of the Incas and much enjoyed in the *Oriente* to this day. These days, *chicha* is brewed from sweet corn, plantains, or manioc. Renaldo, Javier, and I each took a sip (this batch had been brewed from sweet corn) and praised the taste of the brew.

That was no empty compliment. I really do like the taste of *chicha*, although every time I take a sip, I have to push one feature of this drink to the back of my mind. The people making *chicha* need to chew the corn (or plantain or manioc) mush thoroughly and then spit it out again to ensure it turns into an alcoholic drink. Enzymes in saliva break the starch in the corn into sugar molecules that can then ferment to produce alcohol. This chewing and spitting is women's work, but men are the main consumers of *chicha*; life is not always fair in Ecuador's *Oriente*.

After three rounds of the bowl of *chicha* out on the veranda, I thought maybe I could announce my departure. After all, it was getting late. Once again, no one seemed to understand why I was in such a hurry. I had to be hungry after the long journey. "Sure," I said. "Make yourself something to eat. We have lots of food."

Renaldo and Javier shook their heads. Both loved European cuisine, especially Italian pasta with tomato sauce. And that's exactly what they wanted me to cook for them now, please. I realized debating the matter was pointless. I would have to wait a little longer before I could start searching for *Galidacris*. Our hostess led me to her small kitchen, where I channeled my frustration into peeling and chopping onions, garlic, and tomatoes. Who had hired whom? Had I hired the two Siona-Secoya or had they hired me? And what if I didn't find *Galidacris variabilis* here at the end of the world and had to return empty-handed?

All this effort and expense for a tiny grasshopper? All biologists who want to know more about evolution have the same problem. It is almost impossible to observe the processes of evolution in real time because they take so long, way longer than the lifetime of an individual researcher. Ten to one hundred thousand years might pass before a new species of animal or plant develops.

What evolutionary biologists can study, however, are variations within a species. The most obvious are differences in appearance, for example, in coloring—as in the case of *Galidacris variabilis*—or in body size, body hair, and a lot more. The behavior of an animal—and its biochemistry and physiology—can also differ from individual to individual within a species. These differences usually arise from small mutations in the genes of the life-form. These genetic variants are what make evolution possible. Some of these variants are more successful in the great competition for resources, and the altered genes are handed down to succeeding generations; others are disadvantageous, and these whims of nature quickly disappear.

~~~~~~~

SOMETIMES DIFFERENCES WITHIN a species can have both advantages and disadvantages. Migratory birds are a well-known example. Birds in the Northern Hemisphere leave their breeding territories in fall to overwinter in the south, where it is warm. But there are exceptions. Some individuals from various species do not give in to the urge to travel to warmer climes in late fall. Instead, they struggle through winter in the chilly north. Both strategies have advantages and disadvantages. The long flight to overwintering areas is exhausting and fraught with danger, but every harsh winter kills many of the birds that skip the risky journey. In spring, however, the individuals that do not migrate are first in line for prime nesting spots. Every harsh, snowy winter favors the birds that overwinter in the warm south. As the climate changes, such winters are increasingly rare, which means the proportion of the population that does not migrate steadily increases. If climate change continues as feared, the birds that stay put could be the ones that survive, while their migrating cousins will be the ones that die out.

Sometimes, as in the case of the red- and green-legged grasshoppers, it's difficult to tell whether one of the two variants is more advantageous—and, if it is, under what circumstances. What is interesting about *Galidacris variabilis* is that the variations are geographically separated. Ecuador's rainforest river, the Napo, appears to be an almost insurmountable barrier to these little flightless grasshoppers, preventing the red legs and green legs from meeting each other and mating.

The many tributaries to the Amazon River have barely changed course for five million years. Long enough that populations of *Galidacris* separated by the Napo River and unable to exchange genetic material could be drifting apart. The different color variants could be a sign of this. It's likely that the two populations will continue to diverge genetically until *Galidacris variabilis* finally splits into two separate species. Or perhaps the red- and green-legged *Galidacris* hoppers are already two separate species and have been so for a long time.

By definition a species is a community that can reproduce itself. That is to say, when a female and a male mate and produce fertile offspring, they are considered to belong to the same species. If they can't, then they belong to separate species. Big cats exemplify this. If you keep tigers and lions in the same enclosure, they are sometimes attracted to each other. Sometimes the relationship leads to offspring, which are called either ligers or tigons, depending on whether the father was a lion or a tiger. However, these hybrids are infertile, and so lions and tigers are two different species in the genus *Panthera*.

The situation is different if you cross various populations of tigers. Sumatran, Amur, and Bengal tigers would never meet in the wild, but in captivity, they can interbreed and have healthy, fertile offspring. In cases like this, biologists talk of subspecies. If tigers in various parts of Asia survive (and that depends on us), it's conceivable that over thousands of years, a new species could evolve from one of these subspecies, in much the same way a common ancestor once gave rise to lions and tigers.

The grasshopper *Galidacris* indicates to evolutionary biologists that such processes can play out in areas that are right next to each other. A rainforest river can be enough to cause one species to split into two. It's especially interesting that the variation concerns a feature that plays a crucial role in the sexual behavior of this grasshopper. *Galidacris* hoppers wave to each other using their brightly colored hind legs. What if red-legged females simply prefer to mate with red-legged males? Then, even if the Napo River, the barrier between the two grasshopper populations, were to disappear and the red- and green-legged hoppers were to meet up once again, the two populations would remain isolated from each other. Now, the divide would no longer be geographically determined; it would be determined by their sexual preferences. Even if the two separate variants could theoretically still produce offspring together, they would no longer do so.

~~~~~~~~

IT IS SURELY no accident that animal groups such as insects, whose members are especially small, are for the most part more species-rich than, for example, the considerably larger mammals. A river in the rainforest might be a barrier for a tiny grasshopper, but not for a jaguar or a tapir. And so, populations of the tiny representatives of the animal world are more often isolated and develop into different species.

Another reason for the enormous diversity of insects in the tropics is the high number of food specialists that crawl and fly through the forest. There are herbivores that feed on only one or just a few species of plants. Many plants "defend" themselves against insects that want to nibble on them by releasing poisons into their leaves and seeds, which spoil the appetites of most insects. Some insects, in turn, develop resistance to these poisons, overcoming the plants' defense mechanisms. The advantage for the insects is that the previously forbidden fruit is now theirs to enjoy without competition from other herbivores. Populations of specialists develop on different trees and, thanks to their specialized diets, remain isolated from their brothers and sisters on other trees with different defensive mechanisms. Thus, plants become drivers of insect evolution.

The relationship is not a one-way street. The plants then develop new chemical weapons to combat the insects that have adapted to eating them. New plant species develop. Their defensive strategies are cracked. The competition never ends and continually increases species diversity as long as the forest survives.

Eating and being eaten is by no means the only relationship between the different life-forms in a tropical forest. There are the relationships between flowers and pollinators, parasites and hosts, carcasses and carrion eaters, and trees and epiphytes. Insects lay their eggs on plants; birds, bats, and ants spread plants' fruits and seeds. This list is just a small fraction of the wide-ranging web of relationships between animals, plants, fungi, bacteria, and viruses. This web is particularly tightly woven in the tropics. Here, many life-forms have undergone a striking degree of specialization, which means the niches the inhabitants of the tropics claim as their own are correspondingly small, and so more species can live in close proximity and survive.

~~~~~~

MEANWHILE, THE AROMA of tomato sauce was spreading through the home of the Quechua family. We sat out on the veranda and ate the pasta—with a breathtaking view of the Aguarico River. After the last noodle was dispatched and the plates were licked sparklingly clean, I wanted, finally, to leave. But once again, "Wait!"

There was another round of *chicha* to be enjoyed. I should have known. The sun was low in the sky by then, barely over the tops of the trees on the opposite side of the river. The time was running out for any *Galidacris* research today. But here I was sitting with the best rainforest guides imaginable, people who had grown up surrounded by the insects that were of such interest to Klaus and to me.

"Have you ever seen this grasshopper?"

I circulated a photograph that Klaus, the instigator of this expedition, had given me. Our host stood up and disappeared. Less than five minutes later, he was back, holding a large leaf that he had folded to create a small box. Then, as I watched, he unfolded the box and we were looking at an adult male *Galidacris*. Its hind legs were green.

"They're always out there in my garden," our host said casually. "Right behind the house."

That's how easy research into evolution can be.

Listing slightly to one side after our generous sips of *chicha*, we ventured out into a garden covered with the lush growth of manioc and other vegetables. On this late afternoon, we saw more *Galidacris* hoppers. All of them had green legs like our first find. This confirmed what Klaus had suspected: in Ecuador's "Wild East," the resident variant of *Galidacris* is the one with green legs.

—————

AS SO OFTEN happens, however, on closer inspection it all turned out to be much more complicated than that. *Galidacris variabilis* lives up to its name (*variabilis* translates as "changeable") in yet another way. Depending on where they are found, some individuals have shorter or longer white tips to their antennae, and their knees are different colors on both red legs and green legs. It's possible that the species biologists call *Galidacris variabilis* evolved into two or more subspecies a long time ago, and has now become what is known as a polytypic species—a species that contains distinct populations, each with its own defining characteristics.

The take-home message is what the tiny grasshopper's story tells us. It demonstrates how evolution works. And it shows us that the origin of species that Charles Darwin wrote about over 150 years ago continues to this day. Everywhere on Earth, whether in the Amazonian rainforest, on the African savanna, in the beech woods of Europe, or on tropical coral reefs, nature is still adding chapters to this work in progress. Always and everywhere, new species of animals, bacteria, fungi, or plants are evolving.

Unfortunately, it takes much longer for a new grasshopper species to evolve and hop through the rainforest than it does to destroy its rainforest habitat. It took only a few hundred years to cut down most of the cloud forests in the Ecuadorian Andes. Even in the lowlands of the Amazon, roads, cattle ranches, plantations, and settlements are eating their way ever eastward. It is highly probable that some variants of *Galidacris variabilis* disappeared with their habitat before they had the opportunity to evolve into new species.

~~~~~~

LUCKILY, THE WORLD was still as it should be around our hosts' home. The rainforest deep in the *Oriente* was still intact. Now that my assignment was successfully concluded, I began to relax. I sat on the bank of the Aguarico with Renaldo and Javier.

"Sometimes river dolphins swim by here," Renaldo said. "If we're lucky we might see one today." My two companions, however, did not want to wait and got to their feet after a few minutes. "Until later. We've got something we have to do." After a short while, I heard our motorboat chugging down the river.

There is nothing I enjoy more than dusk in the tropics. The heavy heat gives way to a light breeze, cicadas bid goodbye to the day with their twilight concert, and the composition of the rainforest chorus changes. In a place where there's only a handful of people every few miles, living completely without electricity, it wasn't long before an incredible starry sky spread out above me. I took off my shoes, pants, and T-shirt and jumped into the cool waters of the Aguarico. It might sound a bit pathetic, but on only a few occasions in my life have I been happier than I was right then. I was at one with the rainforest, at one with the river, at one with the concert of animal calls, and at one with the gigantic sky sparkling above me. Back on the bank, I let the warm night air dry me and took a bottle of red wine from my pack, a bottle I had brought along for just such a moment as this. No river dolphins swam by, but I knew they had to be out there somewhere.

About an hour later, I heard the chug of the outboard motor once again. My two companions tied the boat up on the bank. Javier told me proudly, "Frank, tomorrow, it's our turn. We're going to cook." Javier held out his arm. From his hand hung a crocodile that had died far too young.

Unfortunately, of all the animals in South America, crocodiles and caimans are the first to disappear. They are easy to hunt and they taste simply amazing. All you have to do is go out after dark in a small boat and drift slowly through the waters where crocodiles live, shining a bright light onto the water. The crocodiles' eyes reflect the light and betray their location like the bright navigation lights on a large tanker. The hunter takes care of the rest with a carefully aimed shot or hard blow with a machete.

~~~~~~

CROCODILES AND PEOPLES coexisted in the Amazon for hundreds of years without any problem. There were simply not enough people living in the rainforest until after the turn of the twentieth century. Things started to get difficult for crocodiles when the rainforest was "discovered" and the oil companies came—and with them, roads and settlers. And—as the young crocodile in Javier's hand attested—biologists traveling to do research could also be responsible for the premature demise of one of these reptiles.

We cannot turn back time. Most Indigenous peoples in the Amazon have been living with refrigerators, cars, motorbikes, money, and hunting rifles with which you can quickly shoot a caiman for a long time now. And yet, in the regions of South

America where rainforests still exist, a new appreciation for the environment has been growing. Many of the Indigenous nations of Ecuador understand how important intact rainforests are for their survival and for the continuation of their culture. In the face of the continued incursion of so-called civilization, caimans, *Galidacris* grasshoppers, and long-tailed macaws have a chance of survival only if large areas of the remaining rainforest are protected.

But even that is no guarantee for all time. Only about sixty miles (about one hundred kilometers) south of the Aguarico River lies Ecuador's largest and, after the Galápagos Islands, most famous national park. The *Parque Nacional Yasuní* is one of the most biodiverse places on Earth, so exceptional that the United Nations has designated it a UNESCO biosphere reserve. Yet even this natural paradise is threatened. The Ecuadorian government has opened up parts of the Yasuní for oil exploration. Civilization, meanwhile, is gnawing at the edges of the Amazonian rainforest from all directions. We can only hope that despite these pressures, we manage to retain the last rainforests—home to unbelievable species diversity and nature's most innovative evolutionary laboratories.

15

Stalking a Spider Cricket

Back in San Pablo de Kantesiya, I focused once again on my own area of research: cricket calls. The search for males advertising for females was somewhat more difficult in the dense rainforest than it had been in the wide-open soccer pitch in the village. Basically, though, I used the same technique. When a single song stood out from the immense rainforest chorus, I directed my microphone toward the sound, put on my headphones, and carefully moved closer. Because I was mostly stalking crickets at night, I wore a small headlamp. If I came too close to the singing insect, it usually stopped abruptly. Then I had to turn off the light until the song started again so I could continue my search.

One night when I was out in the forest, I found myself directly under one of the huge trees. There had to be a relatively large cricket on the trunk somewhere between my knees and my head. At least that was what the loud chirping suggested. I directed my headlamp to the trunk and switched it on. In the circle of light, all I could see was the giant tree's rough brown bark. I took a closer look. Out of the corner of my eye, I caught sight of a shadow scurrying to the other side of the trunk.

Carefully, I followed the nimble creature, and, step by step, I began to circle the tree. The shy insect, however, knew exactly where I was. Before I arrived at the other side, the cricket kept going and arrived back at the spot where this small chase had begun. I changed direction and once again crept around the tree. I struck out again. The little creature was on the move and had disappeared back to the other side. I decided to give it another try and circled the tree for a third time. Yet again the shadow scurried away. We were literally spinning in circles, and I was beginning to take the matter personally.

I directed the beam of my headlamp to the ground and advanced as slowly as I possibly could. That was the right speed for the cautious little insect. After what felt like an eternity, I arrived at its side of the tree. I lifted my head even more slowly, and the circle of light crept up the trunk. I finally spotted the cautious creature. My headlamp was shining on a large cricket with extremely slender legs. Two feelers almost as long as its body stuck out from the end of its abdomen. At first glance, it would be easy to mistake them for another pair of legs. The combination of legs and feelers made the cricket look like a large tarantula. Even longer than the two feelers on its rear end were the two antennae protruding from its forehead—each at least six inches (fifteen centimeters) long and no more than the breadth of a hair. The male cricket was waving its feelers back and forth. It looked as though two fine threads of spider silk were drifting in the breeze. I could see why the insect was so attuned to my movements. With such long, fine feelers, it could pick up the slightest disturbance of air caused by an approaching threat.

A large cricket like this, chirping on a tree trunk without any protective cover, must fear many enemies. Bats with their excellent hearing could locate it and swoop down to pluck it off the trunk. Enormous spiders patrol the trees and grab anything their mouthparts can latch on to. The spiders in Ecuador are real monsters. Some species from the order of the primeval-looking whip spiders grow larger than a human hand. Near their mouths, they have huge appendages armed with spikes that look a lot like the grasping forelegs of praying mantises or the pincers of scorpions. Whip spiders are spider relatives that walk on only six of their eight legs, and the foremost pair don't even look like legs. Over time, these forelegs have developed into sensitive sensory organs even longer than the antennae of the cricket I had just stumbled upon. Whip spiders get their name from these long whiplike legs-*cum*-sensory-organs, which are a superb adaptation for night hunting. Whip spiders use them to detect prey, which they then grab with their terrifying claws.

~~~~~~

CRICKETS' SENSES ARE exceptionally sensitive. The insects have to be constantly on high alert, because they are putting themselves at risk every time they sing. And it's not just large predators that have them in their sights. There are a few species of tiny flies in the family Tachinidae that can hear extremely well. The fly uses its sensitive ears to track down a singing cricket. The minuscule insect then approaches the cricket from behind and tries to land on its victim's abdomen. The only chance the cricket has to escape this aerial attack is to kick out vigorously with its hind legs like a stallion gone wild. If the fly makes it to the cricket's body, the cricket's fate is sealed. The tiny fly deposits freshly hatched larvae onto the cricket. These larvae eat their way into the body of their new host and then gradually consume it from the inside out. After a while, the fully developed larvae emerge from the eviscerated insect and finally turn into tiny flies set on attacking crickets.

If singing is so risky for crickets, why haven't they developed a more discreet way of communicating with each other? Interestingly, the little parasitic flies are the reason some male crickets are falling quiet—at least occasionally. The desert cricket *Gryllus integer* lives in the American Southwest and in northern Mexico. These crickets are often parasitized by a tiny tachinid fly called *Euphasiopteryx ochracea*. The desert cricket has adapted to this threat. It doesn't sing all day and it sings loudest when its little enemy is least active. This lessens the threat posed by the fly, but it does not offer complete protection. Therefore, despite the desert cricket's tactics, many still fall victim to *Euphasiopteryx*.

Some of the desert crickets, however, counter the deadly danger with an even more drastic adaptation: they no longer sing at all. Instead of broadcasting for a female themselves, some males sit silently near a male that is chirping his heart out. This behavior—which is genetically programmed—has earned these silent crickets the moniker "satellite males." They wait until a female is attracted by the singer next to them, and sometimes they succeed in mating with the female in the place of their competitor. This strategy is rewarded with successful mating considerably less often than if the male had undertaken the tedious task of chirping all day long. However, the payoff is that the satellite male is not targeted by tachinid flies and then eaten by their larvae.

The satellite males' ploy works, of course, only as long as there are a sufficient number of singing males they can pair up with. The more parasitic flies live in an area, the more the satellite males benefit from their reproductive strategy. In places where there are fewer flies on the lookout for singing crickets, you find fewer satellite males. *Euphasiopteryx* flies, therefore, have a significant impact on the evolution of desert crickets by directly influencing their behavior.

THE MALE CRICKET on the tree trunk just an arm's length away from me was clearly not a satellite male. After I had remained still, avoiding any hasty movement, for long enough, it lifted its tegmina up a bit from its body and began to broadcast. I let my DAT recorder run for a good minute—my first perfect recording of a cricket calling song in the middle of the forest.

That, however, was the first—and easiest—of my tasks that night. What I had to do next was much more challenging. I had to catch this nimble insect equipped with sensitive sensory organs. The cat-and-mouse game the cricket and I had played earlier was still fresh in my mind.

Animal lovers often do not understand why entomologists have to catch the insects they are researching and then kill them, dry them, and mount them under glass. Unfortunately, there's no avoiding this if we want to grasp the enormous diversity of life in tropical forests. It is only in the laboratory that a biologist can determine with certainty if an insect caught out in the wild is one that has never been described before, or one that has been known to science for a long time. If it is a new species, it's described in a scientific publication, assigned a place within the relationships of an animal family, and given a name. The individual specimen that made it possible for the new species to be described—and that therefore had to die for science—is then stored in one of the many natural history museums around the world as a reference—or type—specimen for biologists in the future. That is why it was so important for my work that I not allow the nimble cricket to escape.

If the little creature jumped off the tree, I would never be able to find it again in the layer of rotting leaves that covered the forest floor. A well-thought-out hunting strategy was therefore required. Moving in slow motion, I put my backpack down, took out a large white mosquito net, and spread it out on the forest floor. Then I took a plastic bag, opened it wide using both hands, and approached the cricket as slowly as I could. When would I be close enough to lunge at the cricket with the bag? I wondered. When I was just a hand's breadth from the insect, I couldn't stand it any longer and I made a grab for it.

I didn't believe it possible, but the insect was quicker than I was. Somehow, at the last moment, it must have slipped around the gaping mouth of the bag speeding toward it and jumped from the tree. I immediately checked the light-colored mosquito net on the ground. I was in luck. The cricket was sitting there waving its long antennae.

Anyone who has been on grasshopper hunts as a child will have quickly discovered that although these insects can jump a long way, in the short moment between jumps they are at a disadvantage. The reason for this is the way their knees are constructed. Before every jump, grasshoppers and crickets have to add tension to their knee joints. It's a bit like winding up the spring in a mechanical toy. They slowly bend their legs, transferring energy from their leg muscles to a cuticle in their knee. When they relax their leg muscles, the energy stored in their knees is released all at once, and the insect bounds away. I absolutely had to stop that from happening. I dropped my plastic bag over the dark shadow on the light-colored mosquito net as quickly as I could. Success! Catch number two bounced around in the plastic bag—catch number one having been the noisy short-tailed cricket from the soccer pitch.

~~~~~

THE NEXT MORNING, I took a closer look at the insect. I had caught a male from the spider cricket family, Phalangopsinae. But whether it was one that had already been described or a new species was something I couldn't confirm from my little laboratory in the rainforest. The analysis wasn't done until two years later by Daniel Otte, a cricket specialist at the Academy of Natural Sciences in Philadelphia. He named the new species *Aclodes cryptos*.[19] Since then, the insect that led me such a dance in the forest around San Pablo de Kantesiya has been part of the academy's collection, and is one of the almost thirty thousand species in the Orthoptera Species File, the scientific database of all the known grasshoppers, locusts, and crickets in the world. *Aclodes cryptos* was my first modest contribution to one of the largest projects in biology: to discover, describe, and record the millions of species of Orthoptera previously unknown to science.

16

Pain on a Scale of 1 to 4

Not every cricket hunt was as successful as my hunt for the spider cricket *Aclodes cryptos*. Some nights, all the nimble singers I laboriously tracked down escaped me. One time, I was stalking a cricket calling from a small bush at the edge of the forest. I moved branches and leaves to the side as carefully as I could and reached in.

An unbelievable pain shot up my thumb. I withdrew my hand immediately. Unfortunately, the burning in my thumb did not become any less painful. Instead, it steadily increased and spread to my whole hand. "Damn! Damn! Damn!" I tried to stem the flood of pain with loud, increasingly vulgar swear words. "Shit! God dammit…"

I shone my flashlight onto my hand. It had taken less than a minute for my thumb to swell to more than twice its normal size. It looked positively grotesque. What really worried me, however, was a droplet of blood welling up out of a minuscule puncture wound. I panicked and ran toward the village—and took the wrong turn at one of the forks in the path. Long after the place where the forest was supposed to have opened up into an overgrown cacao plantation, I was still surrounded by nothing but ancient trees. I stopped and attempted to calm myself down. My thumb was throbbing. The pain kept coming back at full strength at regular intervals, as though whatever demonic creature had bit or stung me was biting or stinging me over and over again. What on earth had done this to me?

I pulled myself together, turned around, and ran back until I found the bush where this had all started. My plan now was to find out what had caused me so much pain. I hoped to confirm what I was beginning to suspect: that I had not been attacked by a small venomous snake but by something relatively benign—despite the fact that the pain this nasty little creature was capable of inflicting was excruciating.

Even people who are not disgusted by insects occasionally get scared and panic. The fear of being stung has spoiled many a breakfast out on the balcony or picnic in the park. Wasps, bees, and most ants sting humans to defend themselves. Some ants bite and then spray venom for the same reason. A sting or a bite from one of these little creatures impresses us so much that after a painful encounter we treat them with great respect. Let's take a moment to consider our fear of insect bites rationally. Instead of being afraid, we should probably be grateful that the feisty Hymenoptera—that is, bees, wasps, and ants—restrict themselves to hurting us instead of killing us outright when we get too close.

~~~~~~~

FEW PEOPLE HAVE immersed themselves so thoroughly in the world of insect stings and the pain they cause than Justin O. Schmidt. Schmidt grew up in the Appalachians, where, even as a teenager, he preferred to take long hikes rather than join the other teenagers at the sports field to watch baseball or football matches. Schmidt was fascinated by insects and left no stone unturned or tree unexamined in his quest to find them. The budding naturalist was especially entranced by the Hymenoptera he encountered on his hikes. Eventually, he became an expert on wasps and ants.

Despite his love of insects, Schmidt became a chemist by profession, not a biologist. But his passion never left him, and he focused his research on the chemistry of the pain insects cause us. Hardly anybody has as much experience with insect stings as he has: on his hikes and travels chasing after new species of Hymenoptera, he has to date personally been bitten or stung by over 150 species of bees, wasps, and ants.

"For a long time, the defensive harvester ants were the ones I respected the most," he told me when I called him up for an interview.

Harvester ants scour wide-open landscapes with little protective cover as they search for plant seeds to feed their brood. That makes them easy prey for birds, lizards, and small mammals. Schmidt is convinced that the persistent pain the ants produce in their victims is an adaptation to their lifestyle. The ants' venom protects them from a multitude of enemies while they are out looking for food.

"What is the most important thing for the little ants?" the chemist asked rhetorically.

"Right," he said, without waiting for my reply. "They want to escape without injury to themselves, if at all possible. It's little help to the ant if it manages to kill its enemy in the end, but the enemy has time to eat or mortally wound it first."

The stab of pain, therefore, must activate as many pain receptors as possible at lightning speed. Only then will the predator leave the ant alone and find itself some other prey. It's even a good thing if the bird or lizard survives this attack. Lots of predators are capable of learning and, after their experience, they give harvester ants a wide berth. And they teach their offspring that harvester ants and other Hymenoptera equipped with effective defensive weaponry are not easy prey.

The more insects stung Schmidt over the years, the better able he was to distinguish the differences between one sting and the next. One species of wasp inflicted moderate pain. Then there were other Hymenoptera whose venom shot through him like an electric shock but quickly faded. The pain from others lasted longer but was more bearable. Unfortunately, the pain that tormented me that night in the rainforest was in a completely different category. It was both extremely intense and extremely long-lasting.

~~~~~~

I DIDN'T HAVE to look for long. Ants were crawling over a branch on the bush—although they were not like any of the European ants I was familiar with. In fact, they were larger than any ant I had ever seen. The ant that had stung me was a conga. This gigantic ant goes by many other names, some inspired by its painful sting: bullet ant and *hormiga 24 horas del dolor* ("ant that inflicts 24 hours of pain") are just a couple of examples. I was experiencing for myself how the nearly inch-long (over two-centimeter-long) ant got names that make it sound like some kind of a weapon. I was, however, also eternally grateful that the creature that had pumped my thumb full of venom was only a conga and not a snake. It meant that although I had many hours of pain ahead of me, I would survive them none the worse for wear.

Schmidt decided to systematically describe and rank the various kinds of pain caused by different species of insects. But how do you express pain in words? The chemist was confronted by the same problem that faces wine and restaurant critics who want to communicate complex culinary experiences to their readers. In the end, Schmidt settled on a solution similar to the ones the testers for Michelin Guides hit upon: he developed a rating system that assigned each insect sting a number, like the number of Michelin stars awarded to fine-dining establishments—only in this case, the number equated to the severity of the pain. But that was not all. Like a wine critic, Schmidt attempted to paint a picture of the particular features of each insect sting in words people could relate to.

The Schmidt Sting Pain Index was born. The scale ranges from 1 for moderately painful experiences to 4+ for the electrifyingly painful stings of the bullet ant, which feel somewhat similar to being shot. In his index, Schmidt describes the sensation unleashed by the giant ant as "pure, intense, brilliant pain—like walking over flaming charcoal with a 3-inch nail in your heel."[20]

I was stung by bullet ants a total of three times during my doctoral research. The pain really is intense and lasts for hours. I never personally experienced the "three-inch nail" piercing my flesh. Perhaps I was just lucky, or received a lower dose of venom than my American colleague. I was, however, reassured by the fact that ant stings, even the most painful ones, don't cause any long-term damage in people.

Despite knowing this, I found the stories told about a Brazilian Indigenous group called the Sateré-Mawé quite incredible. This group lives in the rainforest and bullet ants play an important role in an excruciating initiation rite. Sateré-Mawé men catch the giant ants and sedate them with a special concoction made from plants. Then they make gloves from leaves, weaving the ants into them. When the ants wake up, they are trapped inside the gloves. Young men who want to be recognized as full members of the community must put these gloves on and wear them for several minutes. Participants dance during the ritual, which helps them bear the unbelievable pain inflicted by the ants.

The British naturalist and filmmaker Steve Backshall and his camera team were allowed to take part in one of these bullet ant rituals. Backshall made a film for *National Geographic* in which you can see the young men proudly wearing the ant-filled gloves, hiding their pain to the best of their ability. The participants in the ritual must benefit from the fact that many of the ants probably sting into the plant fibers while the excruciatingly painful gloves are being made, emptying much of the venom in their venom sacs. Nonetheless, when the young men take off the gloves after five to ten minutes, their hands are swollen and remain numb for several hours. Backshall reports that the young Sateré-Mawé men have to undergo this ritual about twenty times. It's astounding what men will do to prove their masculinity.

The pain researcher Justin Schmidt is sometimes accused of secretly being a masochist who enjoys being stung by bees and wasps. He categorically denies this.

"If you work with Hymenoptera, sooner or later you're going to get stung," he explained. "An insect is rarely forced to attack me. A forced sting doesn't reflect what goes on in the real world. In a situation like that, I can never be sure whether an insect injects more or less venom." A forced sting, Schmidt told me, is rarely used in his index.

Schmidt's reputation as a pain connoisseur—or perhaps even enthusiast—might well be due to the wonderfully descriptive quality of the entries in his index. He described the sting of a species of Central American bee (moderate, level-1 pain) with these words: "almost pleasant, a lover just bit your earlobe a little too hard." The African *Megaponera* ant (pain level 2) left Schmidt with the following impression: "the debilitating pain of a migraine contained in the tip of your finger." Other well-known representatives of level 2 include the honeybee and most European wasps. The pain of their stings, too, soon fades and they are harmless, as long as you are not allergic to them.

"It's not surprising," Schmidt continued, "that Hymenoptera have developed stingers and venom as their defensive strategy. Many species form large, sometimes huge, colonies. The members of these colonies drag home enormous amounts of food for the brood in the structures where they live. A bee or ant colony would be a tasty snack for many animals—if the bees, wasps, and ants were not so good at defending themselves."

~~~~~~~

THERE ARE MANY solitary Hymenoptera, however, that are also equipped with formidable stingers. Wasps especially use them less for defense and more often to hunt other insects.

For example, think back to Jean-Henri Fabre's yellow-winged digger wasp: the female uses her stinger and venom to paralyze crickets as food for her young. A relative of the digger wasp, a huge spider wasp, is number two on Schmidt's pain index—at level 4, right behind the bullet ant. *Pepsis grossa* grows to a length of more than two inches (five centimeters). At just over a quarter of an inch (seven millimeters), the female's stinger is clearly visible from a distance. This wasp, like the digger wasp, hunts other arthropods—only, the prey targeted by *Pepsis grossa* is not a harmless cricket. *Pepsis* wasps are large, but the arthropod they target is even larger and heavier: *Pepsis grossa* specializes in hunting tarantulas.

Pepsis grossa lives in the U.S. Southwest, in Central America, and in the northern parts of South America. In the deserts of Arizona and Mexico, it almost exclusively hunts a single species of spider—the enormous Texas brown tarantula. In the U.S., therefore, these wasps are called tarantula hawks. The Texas brown tarantula not only is larger and heavier than the *Pepsis* wasp but also has large, sharp mouthparts and glands filled with a venom that is dangerous to insects. It is an unlikely prey for a wasp. But those are not the only obstacles the hunting *Pepsis* female has to overcome. The Texas brown tarantula spends most of its time safely entrenched in its burrow, and it barricades the entrance with strong threads of spider silk. If the tarantula hawk is going to have a chance in a battle against the tarantula, it must first get the spider out of its burrow. U.S. entomologist Eric R. Eaton watched one of the epic battles between these two predators and described the wasp's strategy.

> Females hunt for their tarantula prey mostly in the morning and evening to avoid overheating in the intense summer sun. Flying low over the ground, they may detect the presence of a tarantula burrow by sight or smell... Once she does find a burrow with a spider inside, [the female] cuts away the silk curtain and cautiously enters the burrow. Soon, both wasp and spider erupt from the burrow. This eviction behavior is crucial to the wasp's success in securing her prey. She would have far less room to maneuver inside the spider's tunnel. The wasp steps back, grooms herself thoroughly, and then sizes up her adversary. She uses her antennae to entice the spider into raising itself off the ground; or even antagonizes the arachnid into a threat posture whereby the tarantula raises its front legs high, exposing its fangs.[21]

This is just what the giant wasp is waiting for, for now she can see the underside of the tarantula's body—the side of the body where the spider's nervous system is located. The wasp attacks and stings the large spider between the base of one of its legs and its sternum. With that, the battle is over: the sting paralyzes the tarantula. The wasp cleans herself and eats a little of the hemolymph (the insect equivalent of blood) now coming out of the puncture wound the wasp's stinger made in the spider's exoskeleton. She usually then drags the helpless spider into its own burrow, lays a single egg, and uses sand to seal the burrow that was once the spider's home and has now been converted into an incubation den for the wasp. You already know what happens next. The wasp larvae eat the paralyzed tarantula alive.

It is fascinating how well every one of the spider wasp's movements fits with the behavior of its seemingly more powerful prey. Even though she is blessed with a tiny brain, she proceeds with incredible precision and overpowers her dangerous adversary—guided only by her instinct to kill.

When *Pepsis* females aren't busy hunting spiders, they are—like the males—quite harmless creatures. They seek out flowers for their nectar and are slow to anger. But once they get riled up, watch out. The tarantula hawk then takes up its defensive position, bending its wings and abdomen—and stinger—forward.

Schmidt has described the pain *Pepsis grossa* can elicit as "blinding, fierce, shockingly electric. A running hair dryer has been dropped into your bubble bath. All you can do is throw yourself on the ground and whimper in pain."

And yet, Schmidt has ranked the sting of the *Pepsis* wasp as easier to bear than that of the bullet ant. The reason for this is that after five to ten minutes, the horrific pain subsides and the worst is over. I also came across tarantula hawks occasionally in Ecuador. I always gave them a wide berth. The sting from the bullet ant was enough for me.

~~~~~~

IN 2015, SCHMIDT was awarded a very special prize for his observations on pain—the Ig Nobel Prize. This slightly satirical prize is meant to honor scientists and scientific achievements that "first make people laugh, and then make them think."

Schmidt is quick to say that his pain index is not meant as a joke. He feels it is important to demonstrate the advantages such a painful weapon offers Hymenoptera in the great struggle for survival. Perhaps the connoisseur of pain has also managed to communicate, at least a little bit, what fascinating creatures insects are, despite—or perhaps precisely because of—the fact that they sting us and give us such a "wonderful" experience of pain. Take the yellowjacket, the common name for a North American species of wasp called *Vespula squamosa*. In Schmidt's words: "Hot and smoky, almost irreverent. Imagine W. C. Fields extinguishing a cigar on your tongue."

17

The Invasion of the Army Ants

Ants are the secret rulers of the rainforest. Thanks to Justin Schmidt's heroic work, we now have a pretty precise knowledge of how these insects' defensive weaponry works. The second important ingredient in the ants' recipe for success is that they form colonies—some of which can be enormous. Ant experts like to see a colony as one gigantic organism. In most cases, all the workers are daughters of a single queen. Only when their ruler produces more queens and male ants, which in turn found new colonies, do the genes of the infertile workers live on.

Worker ants, therefore, have one common goal: to ensure that the colony survives and successfully reproduces. A worker ant will give her all for this goal, including her own life. Once during my time in Ecuador, I experienced for myself just how aggressive and persistent an ant colony can be. An encounter with a huge swarm of marauding army ants cost me two hours of my life and left me with stiff and aching muscles...

THERE ARE A considerably larger number of species of beetles than of ants, but there are more individuals in the ant family than in any other family of insects. Biologists estimate that the total weight of all ants on Earth is more than half that of all other insects combined—far more than the mass of all people alive on Earth today.

Ants have entered into many different types of symbiotic relationships with other animals, fungi, and plants. In the forests of Ecuador, for example, there are trees known as ant trees. These fast-growing denizens of the forest have formed a tight bond with ants in the genus *Azteca*. The trees offer the ant colonies comfortable living quarters, as well as a kind of on-the-house buffet in the form of special nutritious snacks called Müllerian bodies, which provide the ants and their larvae with fat and protein. Moreover, the well-armed ants maintain small herds of aphids, which suck the sugary sap of the ant trees. The aphids are then milked by the ants for their sugary excrement—euphemistically called "honeydew"—much like dairy farmers milking cows for their milk. The tree gets something in return for offering the hungry insects room and board—its ant lodgers protect the tree. It's a bit like a restaurant owner paying protection money to the local mafia. The *Azteca* ants constantly patrol the trunk, branches, and leaves of their ant tree and defend them against leaf-eating pests.

~~~~~~

LEAF-CUTTER ANTS ALSO live in a highly specialized symbiotic relationship with another life-form. Thanks to their uniform diet, they are so completely dependent on a particular fungus that, without it, they would die. The ants, which cut leaves into tiny pieces, are capable of defoliating entire trees. They carry the leaf fragments into their colonies, where millions of ants live underground. The ants cannot digest their cellulose-rich food themselves. Therefore, leaf-cutter ants cultivate fungi in their brood chambers. These fungi break down and digest the leaves for them. At the end of the process, the fungi then become food for the ants. Some species of fungi grow only in the ants' accommodations. Without the continuous stream of worker ants bringing in pieces of leaf to the fungi's underground chambers, the fungi would die, just as the ants would die without the fungi.

~~~~~~~

LEAF-CUTTER ANTS ARE harmless. Bullet ants are painful. Another ant species gives me nightmares in the middle of the day: army ants of the species *Eciton burchellii*. These ants are combative warriors. One colony can contain as many as two million individuals. Just as Genghis Khan's armies swept over the Asian steppes, so *Eciton* soldiers and worker ants sweep through the forests of South and Central America in search of prey. Every last member of an army hundreds of thousands strong is the offspring of a single queen.

In 1997, I was sitting on the veranda at a biological research station belonging to the Pontifical Catholic University of Ecuador (PUCE). The station was in a small reserve in the Ecuadorian Andes called *Reserva Otonga*. Here I was to have an unforgettable encounter with an army of *Eciton* ants.

Giovanni Onore, then professor of biology at PUCE, had bought the small protected area that comprised the *Reserva Otonga* piecemeal, using donations. The forested reserve lies at an elevation of about 6,500 feet (2,000 meters) above sea level. The Italian biologist had invited me to study the crickets in "his" reserve, and I was more than happy to accept his invitation.

The only trails in *Otonga* are narrow and steep. Together with a group of German botanists, I rented a packhorse in the nearest village, San Francisco de las Pampas. We also employed a few villagers to transport our food and equipment up to the biological station in the reserve.

We climbed for about two hours, then entered a forest that seemed almost surreal. For the duration of our stay, we were almost constantly surrounded by clouds. The fog became one with the trees, and the resulting mood was both melancholy and beautiful. The constant high humidity created idyllic conditions for plant growth. Nowhere have I seen so many epiphytes as I saw here.

The trees were encased in a thick layer of moss and hundreds of tightly packed epiphytes sprang from every branch. Each tree looked as though it was cocooned in a green cloak of moss, ferns, lichens, orchids, and countless other plant species—right down to epiphytic cacti.

A sure sign that you are in a South American rainforest is the many bromeliads that flourish on all the tree trunks and branches. These plants occur only in the tropical Americas. In many of the almost three thousand bromeliad species, water accumulates at the base of their leaves, which are arranged in a shape that resembles a drinking cup or chalice. The scientific name for these reservoirs of water is phytotelmata. Thanks to these small pools, South American rainforests are a paradise for frogs, crustaceans, and insects whose larvae live in water: dragonflies, assassin bugs, and thousands of species of flies and mosquitoes.[22]

~~~~~~

COMPARED WITH THE Siona-Secoya forest in the *Oriente* and its many-voiced chorus, *Otonga*'s cloud forest contains very few singing crickets. This is due to the considerably lower temperatures far up the mountain. It is simply too cold in the cloud forest for many species of crickets, with their energy-intensive reproductive behavior.

The forest is home, however, to enormous swarms of *Eciton* army ants. As I just mentioned, you can think of an *Eciton* colony as a mighty army that combs the forest for prey. The female soldier and worker ants form a front as wide as a semitruck is long, and the swarm advances like a walking black carpet. Every insect that flees in the wrong direction (that is to say, into the carpet of ants) is doomed. The *Eciton* army also scales trees as it hunts down anything edible. Tens of thousands of ants rush up the trunk and attack everything that lives up in the crown. It is their sheer numbers that make army ants the rulers of the *Otonga* cloud forest.

You don't even want to think about what might happen if you were to end up in an *Eciton* swarm. Luckily, you get some warning. A swarm of army ants advances on millions of tiny legs, each one of which makes a small noise. Taken together, this adds up to a strange rustling sound. When I first encountered a swarm of millions, I thought the wind was freshening. But as you keep listening, if the swarm is coming your way, the rustling and murmuring continues and, slowly but surely, increases in volume. Eventually the swarm comes into sight: innumerable minuscule individuals crawling over the ground. As the curtain of chitin advances, beetles, moths, and bugs fly up into the air—and the mass killing continues unabated.

~~~~~~~

ONE AFTERNOON, I was sitting on the veranda at the small research station in the reserve. I was alone, because my botanical colleagues were busy with their work in the forest. I was sorting and preparing my collection: dried crickets, the fruits of two weeks' work in the cloud forest. Then I heard a swell of rustling and murmuring, and knew immediately what was moving toward me.

The spectacular behavior of army ants means they have been very well researched. A large number of scientific publications cover their biology, and especially their ecological significance for other life-forms in tropical rainforests. The role they play in the interactions of the forest is truly astounding. Let's stick with the military analogy. Every human army used to be accompanied by a huge number of civilians that traded with the soldiers. Craftspeople, bakers, and butchers all offered goods and services. There were thieves, scribes, battleground artists, translators, and much more. It's very similar with army ants advancing on their prey. They are escorted by an entourage of life-forms looking to profit from the ants.

For over fifty years, U.S. biologists Carl and Marian Rettenmeyer researched army ants and the creatures that followed them as they marched.[23] The scientists sent their own army of undergraduate and postgraduate students to South America over the years. With every expedition, their list of animal species with a direct relationship to army ants grew. Up until his death in 2009, Carl Rettenmeyer collected more and more astonishing information about the ants he studied—after which his wife, Marian, completed his life's work. The couple counted a total of 557 animal species whose fate depended to a greater or lesser extent on *Eciton burchellii*. Over two hundred species of birds were observed around army ant swarms in Central and South America, picking off the insects scared up by the six-legged warriors. Carl Rettenmeyer suspected that twenty-nine of these species are regular followers of the army ants, making them true companion species to *Eciton*. These birds depend completely on the ants doing the laborious work of driving the insects they eat out of their hiding places. Many of these prey species hide in the leaf litter on the forest floor and come out only at night—unless the ants flush them out. Flocks of birds eat their fill as the ant army advances, taking some of the ants' victims for themselves. Behavioral scientists call the ant birds' strategy kleptoparasitism—freeloading through theft.

Army ants are not only accompanied by ant birds. A multitude of butterflies flutter around the ants' feathered followers. The ant birds' droppings are a nutritious, and therefore much-sought-after, resource for the butterflies.

In the vicinity of the ant swarm, spider wasps, relatives of tarantula hawks, are on the lookout for spiders. Parasitic wasps and tiny flies lay their eggs and deposit their larvae on insects that came within a hair's breadth of succumbing to the ant army. Other fly larvae feed off the victims the ants leave in their wake—frogs, for example, which the ants can kill with their sharp mouthparts but cannot cut up and carry away.

~~~~~~

THE BIOLOGICAL RESEARCH station in *Otonga* was a wonderful wooden building supported by eight thick posts about the height of a person. There were large rooms inside, including a dormitory where scientists and students could spend the night. At the time, there were five of us in the reserve. But I was alone as I waited for the ants to attack. I knew that the army ants are not interested in mammals and especially were not interested in me, an adult *Homo sapiens*. The ants are equipped with large cutting mouthparts shaped like shears. These do a great job of chopping hard-scaled insects into bite-sized pieces, but they are no good for cutting into the muscular flesh of mammals. The only things I was afraid for were our provisions and, most of all, the crickets I had so painstakingly collected as part of my thesis research. I therefore decided to confront the advancing army and defend the station against the ant invasion.

I looked around the research station. Were there any weapons I could use to ward off the impending attack? I decided on a wide broom with a long handle and positioned myself where one of the eight posts was attached to the building's floor. I chose the post closest to the advancing ant army. My fears were realized. The army ants climbed the post as though it were an ancient tree. A stream of hard, small bodies welled up through the cracks in the floorboards like water in a flood. I began to sweep as thoroughly and as quickly as I could.

~~~~~~

THE LIST OF life-forms with which *Eciton burchellii* is associated includes a few that are truly tiny. Some creatures take advantage of the fact that army ants build a kind of field camp—a bivouac—in a hollow tree trunk to house their brood and their queen. An *Eciton* colony goes through sedentary stages in the course of the year, during which the ants return home to the same bivouac every evening. The colony also goes through nomadic phases when the worker ants carry the larvae through the forest. Army ants on the move overcome obstacles such as small streams by forming a living bridge with their bodies.

Every evening in a nomadic stage, the huge ant colony sets itself up in a new bivouac. This type of accommodation is a paradise for mites. Some species sit on adult ants like eight-legged pimples and suck their hemolymph. Others parasitize the ant brood. Yet others fall on the waste that the *Eciton* worker ants take out to a kind of trash pile right next to their bivouac. The German ant researcher Stefanie Berghoff once counted the mites that live together with *Eciton burchellii*: about twenty thousand of these little arachnids live as tenants in an army ant colony.

It's also amazing how many species of insects—including a few beetles—manage to live safely in the middle of the bivouac of these insect-eating ants. The beetle species seem to share an unusual behavior. To avoid being eaten by the ants, they camouflage themselves—as ants. They copy the shape and color of their hosts; they also repeatedly touch the ants with their antennae, stealing some of the typical *Eciton* scent for themselves. The beetles rub themselves with the ant perfume and this grooming behavior then masks them chemically as well. Just in case the camouflage and "Eau de *Eciton*" fail them, most of the bivouac-dwelling beetles have an extremely flat, round exoskeleton, perfectly shaped so the mouthparts of the predatory ants cannot gain purchase—and simply slip off.

But why do the defenseless beetles put themselves in such danger by venturing right into the lion's den? The answer is clear. The squatters are living in a land of plenty. A constant stream of worker ants keeps bringing new supplies of food for the ant brood from the forest to the bivouac. All the beetles need to do is help themselves. And that's not all. As long as the beetles' cover is not blown, the army ants' bivouac is one of the safest places for otherwise-defenseless beetles to be. Thousands of soldier ants are defending their brood and, in so doing, are also involuntarily defending the beetles from attack.

~~~~~~

MORE AND MORE ants were streaming out of the cracks between the post and the floor. I speeded up my sweeping to return the creeping critters to whence they had just come. But no matter how hard I swept, the flow of ants just would not stop. I imagined the ants would soon take over the building as my muscles gradually started to burn. But then—relief. The rush of small bodies suddenly slowed, and the ever-diminishing numbers of invaders turned around as soon as they reached the floor. My heroic sweeping seemed to have made a lasting impression.

The most important means of communication for ants is scent. The ants attacked by the broom seemed to have left a chemical warning: "Danger! Squad retreat! Flee!"

Exhausted, I made myself a cup of coffee and sat down for a moment. My colleagues would never believe what I had just been through. But I had let down my guard too soon. A few minutes after I had seen off the first attack, the besieging forces were climbing post number two. Whether I wanted to or not, I had to grab my bristly weapon for a second time and continue sweeping...

OF THE 557 animal species that researchers have observed near *Eciton* colonies, about 300—so Carl Rettenmeyer and his team estimated—are at least partially and temporarily dependent on the ants. The researchers described this as "the largest animal association centered on one species." "The extinction of *E. burchellii* from any habitat over its vast area of distribution is likely to cause the extinction of numerous associated animals at that site," the authors of the study warned.[24]

The swarm of army ants attempted to invade the small research station three more times. Every support post appeared to them to be yet another tree to be conquered for food. Every time I had to sweep like mad or surrender to the million-strong army of ants. At the end of the afternoon, when the last ant had realized that I was not going to give up, I rested against the wall of the building, completely exhausted. I was relieved that the storm had moved on. Luckily, I had brought a bottle of *aguardiente*, Colombian schnapps, along with me to *Reserva Otonga*. I sat down with the bottle in front of the building, took a long swig, and waited for dusk. I couldn't help but think of a sentence from *One Hundred Years of Solitude*, a novel by the Colombian Nobel Prize winner Gabriel García Márquez: "The first of the line is tied to a tree and the last is being eaten by the ants."

So reads the grim prophecy in Márquez's novel, which prefigures the rise and fall of the Buendía family in the rainforest settlement of Macondo. In the end, it is indeed the ants that seal the family's fate when they kill a newborn. I had always understood the gruesome ending of *One Hundred Years of Solitude* to be a fantastic exaggeration and a metaphor for what happens when you give up and let yourself go. In the *Reserva Otonga*, I revisited this thought. Now I understood why people who live in the forests of Ecuador regularly remove even small plants that grow under their dwellings. This lessens the likelihood of uninvited insect guests coming to visit.

～～～～～

THAT EVENING, I sat on the steps of the research station even longer than usual. As though to make it up to me, nature put on an unforgettable spectacle that night. Usually the reserve was swallowed by a cloud every morning and shrouded in thick fog for the remainder of the day. Not so on the evening of the ant attack. The clouds parted, opening up a view to the two craters of the twin Andean volcanoes, the Illinizas.

It'll probably sound like something out of a kitschy Hollywood movie, but on that evening, the first without a curtain of cloud since we had climbed up to the reserve, a full moon rose in the sky. Now, even though night was falling, it was brighter than during the foggy hours of the day. The moonlight woke cicadas and birds that I had otherwise heard only in daytime. Neon-green points of light flitted from tree to tree—fireflies in the mood for love looking for mates. I took another couple of sips of Colombian firewater.

At first tentatively, and then with increasing intensity, the moon changed color, its yellowy gold gradually turning blood red. Was the *aguardiente* working its magic? No. Not only had the curtain of cloud been drawn back, but now a special performance was playing out on the sky's stage: an eclipse of the moon in the starry skies over the Andes.

The botanists were already asleep, likely dreaming of orchids, mosses, and ferns. I took a final sip of *aguardiente* and was happy to be exactly where I was right then.

18

Bidding Adieu to San Pablo

For a whole year, I gathered data for my doctorate in Ecuador, recording cricket calls in two forested areas of the Andes. I spent most of my time, however, in San Pablo de Kantesiya on the Aguarico River in the Amazon Basin, because that was where I was most likely to find what I was looking for. After I analyzed my recordings one thing was clear: the diversity of singing crickets is highest in the lowlands of Ecuador. The higher I went in a study area, the fewer songs there were for me to record. As I had hoped, my soundscape analyses reflected the species diversity of the study area at the time. They confirmed what other entomologists had already determined. The higher you climb in the mountains, the fewer species of insects you discover.

~~~~~~~

IT WAS MOSTLY the small, sometimes tiny, crickets that occurred only in the lowlands of the Amazon and not in the mountains. In the leaf litter on the forest floor, I found miniature crickets about a quarter inch (less than a centimeter) long, with extremely quiet calling songs. Their legs were striped black and white. Unlike most other crickets, these tiny ones were almost impossible to overlook when they crawled over the brown leaves. But it was easy to confuse them with another small creature. At first glance, they looked much like the miniature striped jumping spiders that hunted on the leaf litter. This deception, called mimicry—the ability of some animals to protect themselves by imitating a venomous or well-defended creature— is a common survival strategy in the animal world and is clearly employed by some of the tropical mini-crickets as well.

All crickets undergo a number of different larval stages. Unlike the larvae of butterflies, bees, or flies, cricket larvae resemble adults. They lack the wings their fathers use to sing, but they run on six legs and are clearly identifiable as crickets. They grow larger and larger as they progress from one larval stage to the next. For every adult cricket, there are hundreds more tiny ones. When I moved the rotting leaves on the forest floor to one side, mini-crickets as well as cockroaches in all larval stages scattered, seeking to save themselves. My boss, Klaus Riede, once called this scurrying mass of small insects and their even smaller larvae the "plankton of the forest."

～～～～～

THIS MICROCOSMOS IS the foundation of a massive food web. The word "web" describes the relationship between predators and their prey much better than the more common expression "food chain." The image of a linear chain where a large species eats a slightly smaller one, which eats a middle-sized one that in turn eats a little one that, finally, hunts something tiny is misleading. The relationships in nature—especially in a species-rich rainforest—are considerably more complicated than that.

The mini-insects—the "forest plankton"—are hunted and eaten by spiders, predatory praying mantises, army ants, and hungry frogs and lizards. These, in turn, feed a huge diversity of larger hunters, but the relationships that form the large food web of the rainforest are even more complex than this. Small parasitic flies stalk much larger creatures and wasps attack tarantulas. Mosquitoes and horseflies suck the blood of larger mammals before ending up in the webs of orb-weaver spiders. And that is just a small section of the tightly woven food web.

In this web, it's not only animals with spiny mouthparts, sharp teeth, or pointed beaks that insects such as crickets have to watch out for. Many animals, large and small, fall victim to parasites with wily reproductive strategies. Sometimes the danger is almost invisible: in tropical rainforests, even fungi hunt insects. The cricket only needs to come into contact with the spores of an insect-eating fungus. The fungus grows from the spores that have landed on the insect right into the insect's body, thereby killing its unsuspecting victim. Now the fungus can spread within the chitin carapace and gradually digest the insect. After a while, the fruiting body of the fungus—including the stalk and cap—grows out of the insect's body. A bizarre sight, indeed.

Some predatory fungi literally reprogram the behavior of their prey before they die. In one of their publications, mycologists Harry Evans, Simon Elliot, and David Hughes from Pennsylvania State University describe a group of fungi from the Brazilian rainforest that they called zombie-ant fungi.[25] These predatory fungi invade specific areas of the ants' brains, causing the insects to completely change their behavior. Like zombies in a movie, they now have just one goal. As though remotely controlled, the infected ants leave their nest, crawl up a young plant stem close by, and jam their mouthparts into the underside of a leaf about a foot (thirty centimeters) above the forest floor. The ants die hanging in this position and the fruiting body of the predatory fungus grows out of their heads. The researchers suspect that here, slightly elevated above the ground, the humidity in the air is optimal for the fungus. Ants follow the same pathways into and out of their nests, so this is also the optimal location from which to infect more ants following the pathway along the underside of the leaf to reach their nest. The deadly spores rain down from above on the as-yet-uninfected insects.

~~~~~~~

MY NEIGHBOR RENALDO, the rainforest river boat captain, wanted to share a story with me about a fungus that ate animals.

"You biologists are always looking for medicinal plants. I know a good medicine for when your, ah, your you-know-what, ah, doesn't want to stand up any longer."

Of course, he sparked my curiosity, and I immediately wanted to know more.

"You go out into the forest and kill a toucan with a blowpipe."

Renaldo paused for dramatic effect.

"If it falls on its back, small mushrooms start to grow out of its stomach. You have to eat them."

"And?" I asked. "Did it work for you?"

"Not yet. All the toucans I've killed so far have landed on their stomachs."

~~~~~~~

THE MORE PLANT species a forest is home to, the more leaf- and fruit-eating animals live there. There are also more of the "forest plankton" that live off plant debris. The more components a system like this contains, the more resilient it is to external disruption. The more diverse a habitat is, the less danger, for example, that an introduced animal species will destroy it. In a biodiverse environment, a whole armada of predators and parasites usually stands at the ready to control an "alien" intruder.

A biodiverse habitat is also considerably more resilient in the face of so-called pests. Humans could learn a lesson here. In a rainforest containing hundreds of species of trees, each of which uses a different defensive strategy, plant-eating insects find a source of food they can eat only every hundred yards or so. Moreover, an ancient forest like this offers innumerable niches for insects, birds, reptiles, and mammals, which, in turn, control the pests that eat plants. In a human-planted mono-culture—a spruce plantation, for instance—the situation is quite different. A warm winter followed by an overly dry summer is all it takes, and bark beetles can devastate these monocultures. And there are hardly any predators to keep the enemies of foresters in their place.

Humans should see the extreme species-richness of rain-forests as a model to aspire to. We need to rediscover the value of biodiversity. In some places, this is happening already. Many foresters are beginning to use forest practices that more closely mimic nature. They value species diversity and plant species appropriate for the places where they are being grown. In the end, these practices benefit both nature and forest owners. For-ests that emulate nature are not only more biodiverse but also more resistant to mass destruction by pests. Similar ideas are coming to the fore in agriculture. Permaculture operations are aiming for healthy ecosystems. People who follow the prac-tices of permaculture raise many different sorts of fruits and vegetables and livestock in close proximity to each other with resounding success. Permaculture gardens offer many niches for plants and animals. "Beneficials" keep "pests" in check. Yields from permaculture operations can therefore be high, even when pesticides are not used.

BUT BACK TO the diversity of crickets in the forest around San Pablo de Kantesiya. Species of crickets that sang on the forest floor and low down on trunks were easy for me to catch. After a while, however, I stopped coming across species that were new to me, no matter how hard I looked. It seemed I had collected almost all the cricket fauna that lived on the forest floor. I was now concerned about the songs ringing down from the crowns of the trees some eighty feet (twenty-four meters) above the ground. How was I going to catch the crickets that lived up there?

I approached the problem from various angles. With Javier, who had accompanied me in my quest for *Galidacris*, I looked for a suitable place from which to reach the rainforest canopy. When we found two trees growing barely three feet (one meter) apart, we transformed the two trunks into a ladder by nailing sturdy branches across the trunks to use as rungs. Up in the canopy, Javier put together a platform where I could work.

When evening came, I climbed up to my research platform sixty-five feet (twenty meters) above the rainforest floor. As night fell, the cricket chorus began. I succeeded in getting wonderful recordings of many canopy residents, but the singers I heard remained beyond my reach. My platform was much too small.

I had decided to give up, climb down, and walk home, when there was a crackling and rustling in the branches at the edge of the canopy. I turned on my headlamp. Two yellow gemstones sparkled at me in the lamplight—the oversized eyes of a curious nocturnal monkey were staring at me and reflecting the light from my headlamp. The animal was not at all shy and came closer to inspect the strange construction work in its territory. My night visitor remained calm and unconcerned. After twenty, thirty seconds, during which we stared at each other, the monkey disappeared into the darkness. I decided to stay a while longer.

Suddenly, there was a bright flash of light. Fear shot through me. To whom did this yellowy-green flashlight belong? I hoped it was not an armed Siona-Secoya hunting for monkeys. I turned on my headlamp and cleared my throat loudly to make it clear I was human.

The light flew directly at me.

A luminous beetle the length of my finger landed right next to me, attracted by the light from my headlamp. This was not like any firefly I knew from Europe. The large insect shone its yellowish light from three light organs, two on its back and another one on its stomach. The beetle now in front of me was a record holder. *Pyrophorus noctilucus*, a click beetle, is listed in the University of Florida Book of Insect Records[26] as the brightest bioluminescent insect of them all. Unlike many tropical fireflies, which have light organs that blink, the light from *Pyrophorus* beetles shines continuously. Having three light sources allows *Pyrophorus* to shine in all directions. But it can do even more than that: it can also control how brightly its lights shine. If a potential enemy is approaching, the beetle can activate an internal switch that turns up the intensity of its bioluminescence. *Pyrophorus* probably becomes brighter to signal to predators such as bats that they better think twice before grabbing them. Click beetles don't make a tasty night-time snack. Indeed, they contain foul-tasting substances that make them a bitter pill to swallow.

~~~~~~

LIGHT ALSO TURNED out to be the solution to my own problems getting close to the crickets that hung out in the canopy. I had brought with me from Germany a long narrow lamp shaped like a rod. In Poza Honda, the tiny trading post on the Aguarico River, I'd bought a car battery. Every evening, I lugged my new power source into the rainforest, heaved it up onto my platform, and attached the light to the two poles of the battery. The rod glowed with a blue light and as I held it, it reminded me of a lightsaber from the Star Wars movies.

I hung a white bedsheet between two branches and I shined my light on it. All I had to do now was wait. Gradually, more and more night-flying insects flew in. The beam of light mostly attracted moths: from tiny ones to emperor moths with eight-inch (over twenty-centimeter) wingspans. But there were also giant flying cockroaches, nocturnal wasps, and cicadas. And, luckily for me, flying crickets. I gathered up the crickets, dumped them into plastic containers with perforated lids, and took them back to my hut. I lined the containers up next to my sleeping spot. Then, when I went to bed, I put my recorder next to my pillow.

Every once in a while, I was woken by the chirp of a cricket. "*Criii, criii, criii.*"

I would feel for the recorder, put on my headphones, and start recording. I pointed the microphone at one container after another.

"*Criii, criii, criii.*"

When the microphone was pointed at the container that contained the singing cricket, the sound in my headphones got louder. I was always happy when I came across a new song—and a new chapter for my thesis.

Over time, I identified over forty different species of crickets from their songs in a research area about the size of two or three soccer pitches—including species that were louder and even more disruptive of sleep than the *Anurogryllus* I had found on the community playing field.

Most species were small, sang as clear as a bell, and lived in the continuous canopy of the rainforest. Trigonidiinae, or sword-tail crickets, are the most species-rich subfamily of crickets in the Amazonian lowlands. They get their name from the adult females, which have a saber-shaped extension on their abdomen. The females use this extension to pierce soft tissue in trees, bushes, and epiphytes. The female then lays her eggs in the holes she has made. Unlike most other crickets, which deposit their eggs in holes bored in the moist forest floor, sword-tail crickets are perfectly adapted to life in the tree canopy.

~~~~~~

OF COURSE, EVERY once in a while, I asked myself if I might have spent months in Ecuador's rainforest to help gather useless knowledge, but, on balance, I think not. I discovered a few new species and genera of crickets. One, a spider cricket, was even named in honor of my doctoral supervisor, Martin Dambach. It is called *Dambachia eritheles*.

Okay, so a couple of new species of crickets aren't that special and are of interest to just a handful of scientists. What is of more importance, however, is another result from my research. It turns out that even though so many crickets live together in this section of the rainforest, basically no song that is typical for a given species is identical to the song of any other species. The multitude of singing crickets divide their acoustic communication channels up among themselves perfectly. Their calling songs are differentiated by the frequency of the calls or the rhythm of the pulses of sound. Every species occupies a unique acoustic niche.

And yet, as so often in this world, here too there is an exception to the rule. The calling songs of two species of sword-tail crickets are indistinguishable. Each species uses the identical frequency and raises and lowers their wings at the exact same rate. But one is active at night, while the other advertises for females during the day. Using temporal niches, both species avoid misunderstandings when attracting sexual partners.

They are never on the stage of the great rainforest opera house at the same time.

The crickets in the rainforest around San Pablo prove what other experiments in tropical rainforests have shown. The species that live in these habitats are extremely specialized and the niches they occupy are sometimes tiny. This is how the enormous number of species divide available resources, whether the resource in question is food or bandwidth.

Nowhere in the world will you hear a more diverse chorus than in tropical rainforests. I only hope that many more generations of people will have the opportunity to eavesdrop on a performance staged by the gigantic jungle opera.

~~~~~~

AFTER I HAD traveled back and forth between Germany and Ecuador three times, and after dozens of nights chasing crickets in the forest around San Pablo, it was time for me to say goodbye to the Siona-Secoya. I gave away my mosquito net, my car battery, my tools, and my malaria tablets, packed up my traps, and embarked on my journey back to the bone-chilling damp of the German city of Cologne in November.

Today, twenty-five years later, you can travel to the rainforest village of San Pablo de Kantesiya by road and, I'm told, lots of the people who live there ride around on dirt bikes. Why not? *Indígenas* should be able to do the same things people in other parts of the world do. I, however, would prefer to keep my memories of the village in the rainforest intact. I have returned to Ecuador a few times, but I have never been back to San Pablo de Kantesiya and the magical Aguarico River.

PART III

Giving Nature a Fighting Chance

19

The Land Falls Silent

After I finished my doctorate, I had a difficult decision to make. Should I stay at the university or dare to enter the real world outside of academia? In the end, I became a journalist and documentary filmmaker. I was lucky. Within the media world, I found a niche where I could stay true to the themes that had held my interest since childhood: my love of nature and my interest in animals and their behavior. This seemed fitting, as it had been television that awoke my passion for the natural world in the first place. The program *Ein Platz für Tiere* (*A Place for Animals*) by Bernhard Grzimek is one of the first television shows I remember. In my mind's eye, I often accompanied the biologist, zoo director, and Academy Award–winning documentary filmmaker as he flew over the Serengeti to count gnus and zebras.

When I got older, I shifted my allegiance to the eco-journalist Horst Stern. I have never forgotten a film he made with the former director of the Cologne Zoo, Ernst Kullmann, which got people to see a particular group of animals in a more positive light.

Stern had a television series called *Sterns Stunde* (*Stern's Hour*). In 1975, I was sitting in front of the television with the rest of my family when an episode of this show came on called "Leben am seidenen Faden" ("Life on a Silken Thread"). My mother and I could not take our eyes off the screen. Stern was a genius doing something no one had ever dared to do before: he narrated a nature program all about spiders. Not a single adorable feathered creature or soft furry baby or majestic lion or elephant appeared.

Stern and Kullmann used cutting-edge cinematography to present their eight-legged protagonists in ways never before seen. The web-spinning and jumping spiders filled the screen of our family's very first color television. Some species close up looked more like lovable clowns with four gigantic eyes than the living nightmares conjured up by arachnophobes. The images were accompanied by remarkable stories. We learned that there are water spiders that hunt aquatic arthropods in German ponds. They use their spider silk to spin little diving bells so they can catch and eat their prey underwater. Stern and Kullmann filmed a male nursery web spider, *Pisaura mirabilis*, catching a fly and offering it to a considerably larger female as a nuptial gift—something he needed to do if he was to survive mating with her. The gift ensured that his chosen one, an eight-legged *femme fatale*, would then eat the fly and not her suitor during the act.

"Life on a Silken Thread" had a huge impact on our family and completely changed the way we viewed spiders. Before that evening in front of the television, any spider that wandered into our home was unceremoniously sucked into the vacuum cleaner. After watching Stern's masterpiece, my mother scooped up every creeping creature that came to visit—whether it was a spider or an insect—and carried it in her bare hands out over the terrace and into the backyard. That episode of *Stern's Hour* really was an hour of reckoning in our lives.

I found a 1996 French documentary by Marie Pérennou and Claude Nuridsany equally inspiring. The action in *Microcosmos: People of the Grass* took place in a meadow. In this production, too, unbelievably beautiful macro shots turned tiny arthropods into fascinating characters: ants that milked aphids, caterpillars that munched in unison on leaves, and beetles bombarded and shaken by raindrops. These two filmmakers told stories with their high-tech cameras that echoed the stories Jean-Henri Fabre had so eloquently put to paper about a hundred years earlier. Except—and this was what was so amazing—*Microcosmos* was a film without words. There was no commentary at all. Instead, all the viewer had to do was sit back and enjoy a summer's day while the film revealed what was crawling on the ground around them and fluttering through the air above their heads.

And so, I had examples to inspire me, and after a few detours, I was now starting to make my own television programs about science, the environment, and my favorite subject—animals.

~~~~~~

AS I DROVE through the Upper Rhine Valley in the summer of 2018 while working on a television documentary, fields of maize—and nothing but maize—stretched out on either side of the road. This is what allegedly picturesque Alsace looks like today: an enormous, unending monoculture of maize—to be used as fuel for biogas plants and feed for factory farms full of pigs.

If we had a time machine and could travel back just sixty years, we'd find ourselves in a completely different landscape, one in which many often-overlooked animals were much better off. The crops were left to stand in the fields longer, providing cover and food for small mammals, and fields were edged with strips of meadow that offered food and protection once the crops were harvested. The meadows were also places where butterflies could sip their fill of nectar, countless numbers of native bees could gather food, and birds could snatch up insects for their young. The countryside back then was a varied mosaic of grain and vegetable fields, pastures, hedges, and fallow land. Today's agrarian landscape is more like a large uninterrupted plain growing nothing but members of the grass family. Skylarks used to fly by the thousands over the fields in Europe singing their beautiful songs; today, they are hardly ever heard. The harrier, a kind of hawk, used to hunt in these fields, and hares used to raise their young in them.

Recent times have shown that it is not only species such as rhinoceroses, tigers, and orangutans that are in danger. We are in the middle of an enormous wave of extinction across almost every animal class. Here are just a few examples: Amphibians around the world are dying of fungal infections that people are bringing to the remotest parts of the Earth. Species imported to areas where they are not native are threatening the survival of native species—for instance, American crayfish imported into Europe are pushing out almost all the European freshwater crayfish. In North America, zebra mussels that arrived in the ballast of ships from Eurasia have spread rapidly in the Great Lakes region, outcompeting native mussels. The number of birds around the world, even species that were once widespread, has been declining drastically for years. Seventy of the estimated five hundred species of sharks worldwide are threatened with extinction because dozens of millions of these predatory fish end up as bycatch in the nets of trawlers. Ninety percent of the forests in Madagascar have been cut down, threatening the future of the unique fauna on the island.

Twenty-three of the one hundred species of lemur—prosimians that live only in Madagascar—are at least severely threatened if not close to extinction. In May 2019, the UN organization IPBES—the Intergovernmental Science-Policy Platform on Biodiversity and Ecosystem Services—sounded the alarm. An estimated half a million to one million species are facing extinction. The main reasons are agriculture, deforestation, fisheries, hunting, and climate change.[27] It would be easy to add to this list. And, of course, the large die-off also affects the most biodiverse group of animals of all—beetles, true bugs, grasshoppers, bees, and dragonflies. In other words, insects.

IN 1962, RACHEL Carson wrote a book that startled a generation and became a kind of bible for the environmental movement. The book moved many readers, and some things did change for the better. And yet, Rachel Carson's most dire predictions are becoming reality in some places today, despite all we now know about environmental issues.

In her bestseller, *Silent Spring*, Carson paints a picture of a town in America in the near future where people completely subjugate nature, overuse pesticides, and wipe out everything living in the wild. It is a silent world, a world without any animal voices:

> Everywhere was a shadow of death... There was a strange stillness. The birds, for example—where had they gone? Many people spoke of them, puzzled and disturbed. The feeding stations in the backyards were deserted. The few birds seen anywhere were moribund; they trembled violently and could not fly. It was a spring without voices. On the mornings that had once throbbed with the dawn chorus of robins, catbirds, doves, jays, wrens, and scores of

other bird voices there was now no sound; only silence lay over the fields and woods and marsh.[28]

From today's perspective, Rachel Carson's warning might sound a little like a corny, overly romantic depiction of doom. But we mustn't forget that in the early 1960s, we didn't know nearly as much about ecological connections as we do today. Rachel Carson wanted to get this information out to a broader public—and she wanted to change things. The massive use of pesticides in agriculture was worrying the biologist and author, especially the use of dichlorodiphenyltrichloroethane, or DDT for short.

Research by ornithologists suggested that DDT was responsible for the sharp decline in some bird species. Where DDT was sprayed, birds either died directly or became infertile and no longer reproduced. Traces of the insecticide were found in the ovaries of females and the testes of males. Bird die-off due to DDT impacted many species—from majestic bald eagles, the emblem of the United States, to the American robin, a pretty thrush-sized bird with a red breast that spends most of its time searching the ground for earthworms.

As few as 11 large earthworms can transfer a lethal dose of DDT to a robin. And 11 worms form a small part of a day's rations to a bird that eats 10 to 12 earthworms in as many minutes.[29]

Carson's warning met with success. DDT was banned as an insecticide. And not only that—in the U.S. and Europe, the environmental movement was born. And many things really did improve. From today's perspective, for example, it seems almost unbelievable that there were no biological treatment processes in most sewage treatment plants until the 1970s. Rivers everywhere were inundated with nutrient-rich feces from human toilets. Rivers became sewers brimming with excess fertilizer. Every summer, the Rhine, the river of my youth, seemed close to the point of no return. Thanks to all that fertilizer, algae bloomed. When the algae died, they accumulated on the river bottom. Decomposing algae depleted the river of oxygen, and fish and small aquatic organisms suffocated. The Rhine was nothing more than a disgusting open sewer. This degradation of the river was not only stopped but reversed. Although the situation is not perfect today, thanks to modern treatment methods, waterways all over Germany are now doing much better than they were back in my childhood, when it seemed impossible that one would ever dip so much as a toe into the stinking brew that passed for a river.

The ban on DDT also gave the birds a reprieve. Moreover, increasing ecological awareness meant that nature conservation areas were established in many places, and animal and plant habitats were preserved even in industrialized countries. Many things seemed to be headed in the right direction.

<hr />

AND YET, RACHEL Carson was, unfortunately, correct. Some places have fallen awfully silent.

The skylark, for instance, was one of the most common birds in Germany for decades. In spring, you could hear male skylarks singing loudly as they soared over fields, heathlands, and pastures. They dove repeatedly as they sang in a fantastic mating display that rang in spring all over the country. Since then, however, this once-common bird has been disappearing from the German landscape. And as it leaves, spring is becoming quieter—just as Rachel Carson foretold. Between 1990 and 2015, the number of skylarks in Germany fell by 38 percent. More than every third mating pair simply disappeared from nature's stage.

The skylark is not alone. It is just one example of a general trend. It is, first and foremost, the species of birds that had the most individuals whose numbers are plummeting. This was made clear in a study published by British researcher Richard Inger in 2015 that reviewed counts of 520 species of birds from fifty-two countries in Europe.[30] The astonishing result: numbers of once-common species were tanking, while the numbers of a few rare birds were, at least in some places, increasing so rapidly one might even talk of a comeback. How could that be?

Richard Inger suspected that the rare species were benefiting from concrete conservation measures. The common species, in contrast, were being adversely affected by what was happening on a large scale in agricultural regions. Hedgerows and meadows full of wildflowers were giving way to enormous monocultures. The landscape of industrial-scale farming offered no spaces where skylarks and lapwings could thrive.

Every year, more and more studies confirm this alarming trend. A team from the Senckenberg Biodiversity and Climate Research Centre in Frankfurt, headed by Diana Bowler,[31] evaluated data from bird counts all over Europe. There were catastrophic declines in the numbers of grasshopper warblers, wheatears, and meadow pipits. All three species eat insects and prefer to live in open grassland—a habitat that is becoming increasingly rare in agricultural areas. Two developments hit these species especially hard: habitat loss and the disappearance of the food they rely on—insects.

～～～～～

SINCE 1970, MORE than a third of all species-rich wetlands around the world have been lost, as rivers have been redirected and marshes and tributaries have been drained to create fields, pastures, and agricultural land. Until the nineteenth century, for example, the Upper Rhine Valley, between

Basel in Switzerland and Karlsruhe in Germany, was an enormous labyrinth of meandering streams, tributaries, marshes, seasonally flooded meadows, and reedbeds. Life for people was hard. Every year, they were threatened by devastating floods. Swamp fever, another name for malaria, was rampant in the valley between the Black Forest in Germany and the Vosges mountains in France. People probably thought it was a huge improvement when engineers moved in and forced the Rhine into a channel just 650 feet (200 meters) wide. The river became navigable, powerful hydroelectric plants generated electricity, and settlements expanded. There's no question that the whole region blossomed economically. For nature, however, these supposed improvements were an unmitigated disaster.

Only a fraction of the wetlands once so common in this region remain. Practically every patch of land has been exploited in one way or another. In summer, vast monocultures—mostly of maize, plant after plant, one right next to the other—stretch as far as the eye can see.

In spring and summer, what a humming and buzzing must have once filled the air along the streams trickling down to the Rhine—and in marshes, flowering meadows, and small pastures scattered with fruit trees—as dragonflies, caddis flies, mayflies, mosquitoes, horseflies, grasshoppers, and butterflies all provided ample food for countless species of birds, amphibians, and fish. Back then, no one counted how many insects flew and crawled through the landscape, which makes it impossible to quantify how many species of insects we have already lost.

It is only since the 1980s that we've had reliable data about the biomass of insects. A publication by amateur entomologists in Krefeld in northwest Germany hit academia and media outlets around the world like a lightning bolt in 2013. Members of the Krefeld Entomological Society had set up what are known as Malaise traps in two places in a nature preserve in Krefeld back in 1989. Insects fly horizontally into nets strung up like tents. The insects try to escape by flying up to the top, where they fall into a collection jar filled with a solution of alcohol. Malaise traps, named after the Swedish entomologist René Malaise who invented them in the 1930s, are commonly used to collect flying insects in the field. Between May and October 1989, the amateur entomologists in Krefeld collected about 5.5 pounds (2.5 kilograms) of insects in their two traps.

They repeated their experiment in 2013. The volunteers were careful to follow as exactly as they could all the protocols used twenty-four years earlier. They used the same traps and placed them in the same spots. They emptied the traps and weighed their catch in just the same way as before. The results were frightening. In 2013, they caught only about 19 ounces (540 grams) of insects in both Malaise traps, a reduction of over 75 percent—and that was in the middle of a nature preserve. They also noticed that insectivorous birds such as the red-backed shrike had disappeared from their study area. They summarized their alarming results at the end of their study with these sobering words:

> In our view, it is not possible, on the basis of the measured variables, to analyze the root cause of the comparative biomasses being considered here. With regard to the effect, however, it can be assumed that the above-described reduction in the biomass of flight-active insects on this scale has serious consequences for local biodiversity, food webs, and essential ecosystem functions.[32]

In other words, reality has suddenly caught up with Rachel Carson's fear of a "silent spring." Alarmed by the findings of their colleagues in Krefeld, many entomologists began to take a closer look at their own experiments. Unfortunately, the results from Krefeld did not seem to be an anomaly. A number of publications confirmed fears that the rate at which insects are dying is extremely worrying.

~~~~~~~

IN 2019, TWO biologists, Francisco Sánchez-Bayo and Kris A. G. Wyckhuys, reviewed seventy-three studies of insect die-off from around the world.[33] Their conclusions are downright apocalyptic. Even in places where people rarely travel, the biomass of insects is steadily declining—at a rate of about 2.8 percent a year. The two biologists fear that 40 percent of all insects could die out completely in the next few decades. Hundreds of thousands of species could simply disappear.

When you consider that insects are food for reptiles, birds, frogs, and even fungi, and that many plants depend on insects for pollination, insect die-off can be seen as catastrophic. If we don't manage to change course, whole ecosystems could collapse.

It is also deeply unsettling that insects are disappearing from more than just landscapes reshaped by humans. A similar story is playing out even in a seemingly undisturbed forest in Costa Rica: over 70 percent fewer arthropods in twenty-six years. The effects on the food web within this forest are already visible. The populations of birds, frogs, and lizards are declining in lockstep with those of insects.[34] In tropical forests, climate change is likely to blame for insect die-off. Most studies, however, lay the blame for the approaching catastrophe squarely on habitat loss and increasingly intensive agricultural practices, including the use of pesticides.

~~~~~~~

I ADMIT THAT it's easy to lose hope when confronted with all these horror stories. You might feel that everything is already lost and then completely give up and just wait for the worst to happen. Luckily, however, there are people who refuse to do that and instead fight to preserve biodiversity. That is the good news: sometimes relatively small steps make a big difference when it comes to countering the great extinction of species.

20

Of Spectacled Bears, Antpittas, and Little Stinkers

S ometimes it seems as though there is no hope. Birds
and insects are in rapid decline, virgin forests are being
razed, the climate is changing. The ecological and economic challenges facing us seem insurmountable.

And yet, many of us would like nothing better than to roll
up our sleeves and give something back to nature. I never cease
to be impressed when I meet people who are doing just that.
People like this exist; most are idealists who also recognize
the realities of the world we live in today. On my latest trip to
Ecuador—to make a documentary film about the last remaining cloud forests in the Andes—I got to know quite a few of
these eco-heroes, including a whole village of conservationists
and two brothers who, like St. Francis of Assisi, talk to birds.

~~~~~~

THE IDEA OF the equatorial Andes as an impenetrable wilderness is outdated. Indeed, these mountains are among the

most densely settled areas in Ecuador. Long before the Spanish conquered and settled South America, the Inca and other Indigenous peoples were making inroads into many of the forests growing on the mountain slopes. In recent times, the rate of settlement has speeded up. Everywhere, the forest has ceded space to fields, towns, and roads. If you travel the famous *Ruta Panamericana*, the north-south connector from North America to Tierra del Fuego, you often get the impression you are traveling through the gently rolling foothills of the Alps. Right and left stretch soft green undulating pastures. Cows are grazing everywhere you look, many of them black-and-white Holsteins. It's all very reminiscent of Europe. It's easy to forget you are 6,500 feet (2,000 meters) above sea level—and what look like gentle hills are in fact really high mountains. In all directions, the peaks of snow-covered volcanoes tower above this peaceful scene, some of them more than 16,500 feet (5,000 meters) high. In the middle of this landscape lies a little village of just a couple of dozen houses: Yunguilla.

I got to know Yunguilla thanks to an ecologist from Ecuador's capital, Quito.

"You just have to get to know this place," Santiago Molina enthused. "I have completely fallen in love with the community there."

To completely understand the small miracle that is taking place in Yunguilla, you first have to know that the tiny village lies less than an hour's drive from the outskirts of Quito. Quito, which at 9,350 feet (2,850 meters) is the highest capital city in the world (other than perhaps La Paz, one of two much-disputed capitals in Bolivia), has a population of over two million. Where once dense mountain forests grew, houses now sprout up out of the ground, snaking for thirty miles (roughly fifty kilometers) down a long narrow mountain valley like a huge reptile made of concrete.

Santiago and I left the city by car early in the morning and crossed the equator north of Quito. After ten or maybe fifteen miles or so (twenty to thirty kilometers), my companion instructed me to turn off onto a bumpy side road. A few minutes later, we parked in front of a wooden house—the little village shop in Yunguilla.

I had not seen houses nearly as beautiful as the ones in Yunguilla anywhere else in Ecuador. There was a whole row of tasteful buildings constructed from stone and wood—quite different from most of the poor Andean villages. By Ecuadorian standards, the place looked more than prosperous.

Santiago introduced me to a few of the villagers. Gradually I began to understand what made Yunguilla different from most other communities in the Andes—and it was a real recipe for success. The farmers had learned to live with an animal that many farmers in other parts of Ecuador hate with a passion. Although this animal is strictly protected and those who harm it face severe penalties, people in the Andes still shoot this animal, even today.

~~~~~~

THE ANIMAL THAT has turned life in Yunguilla upside down is the Andean short-faced bear. Because many of these large mammals have patches of lighter fur around their eyes, which makes them look as though they are wearing glasses, they are also called spectacled bears. An estimated eighteen thousand spectacled bears still live in South America, with about two thousand living in the small country of Ecuador. For a long time, spectacled bears were mercilessly hunted, not only for their meat, but also because they occasionally develop a taste for calves and even adult cows. Mostly, however, these mammals in the order Carnivora follow a vegetarian diet. They undertake long journeys from the cloud forests to open meadows above the tree line at an elevation of almost 16,500 feet (5,000 meters) in search of nutritious berries and the fleshy hearts of bromeliads. In most areas beyond the boundaries of national parks and reserves, the bears have to share their habitat with farmers and their cows, as undisturbed forests and areas of open alpine tundra—the *páramos*—are becoming increasingly rare.

Sometimes cows stumble on the steep Andean pastures, and sometimes they injure themselves so severely that they die. A carcass is free food for a hungry Andean bear. Once bears have developed a taste for beef, a few may start to go after living cattle. It's little wonder, then, that many farmers in South America are less than happy to share their pastures with bears.

~~~~~~

ALMOST ALL THE people who live in Yunguilla are farmers. And yet, for nearly twenty years, most families in the neighborhood have struck a sort of deal. Instead of contributing to the destruction of the nature around them, they want to profit from it—and from the spectacled bears.

It all began with a couple of nature enthusiasts who wandered into Yunguilla via an old smugglers' route, a narrow trail, or *culunco*, that led from a low-lying forest reserve high up into the mountains and on to Yunguilla. Yunguilla at that time didn't have much to offer tourists. It had no hot springs or waterfalls. There was just Yunguilla, a small peaceful village full of friendly people.

Twenty families in the small community got together and formed a cooperative. Their goal was to tempt more tourists to come to Yunguilla. But they wanted to try something special and different, and they proceeded cautiously with their plan. Instead of building a hotel, people beautified their homes and offered accommodation under their own roofs. Working together, the families built a creamery to make cheese, a small facility to make jam, and a restaurant with a spectacular view of the surrounding Andean peaks. Visitors were invited to come and simply partake of village life.

Because bears and pumas were often sighted near the village, the ecologist Santiago Molina became interested in Yunguilla. He wanted to find out how many of the rare bears were wandering around just outside the metropolis of Quito, and he enlisted a few villagers to assist him. It's not easy to track down these shy bears. Luckily, modern technology can help. Santiago and his eco-friendly assistants installed wildlife cameras in the forests around Yunguilla. These cameras automatically take a picture when something moves in front of the lens. Santiago and his crew also rammed a stick into the ground within range of each camera. To this stick, they tied a cloth infused with a scent these large furry mammals find irresistible.

"We use an extract of tears of a rare species of bat," Santiago explained to me with a deadpan expression. "A couple of the boys in the village have to keep climbing up the trees to collect them."

I stared at him incredulously for a couple of seconds. Then the penny dropped, and I realized he was having me on.

"No, no," he grinned. "Vanilla perfume that costs a couple of cents at the drugstore is what works best to attract bears and pumas to our cameras."

The photographs the cameras took surprised everyone and absolutely thrilled Santiago Molina. The forest was full of deer, predatory cats from ocelots to pumas—and bears, lots of bears, more bears than Santiago had dreamed were possible. Using the photographs, Molina and his assistants identified sixty individual bears from their facial markings. So many large mammals just a hop, skip, and a jump from Quito.

It was, however, thanks to one very special spectacled bear that Yunguilla became known as the "bear village," Molina reminisced.

"One day when I was in Quito, the phone rang," he told me. "It was a call for help from Yunguilla."

A farmer had found a bear cub that had obviously been abandoned by its mother. Flies had laid eggs in its skin—a sure sign the mother was no longer caring for her baby. Maybe she had died. The bear cub looked as if it might die soon, too. The call was to ask what could be done for the helpless animal.

"I asked if there was a woman in the village who was nursing a baby. If there was, perhaps they could persuade her to give some of her milk to the bear cub. And they did indeed find a nursing mother who agreed to save the baby bear from starvation."

And that was how the bear cub, which the villagers called Yumbo, survived until Santiago Molina picked him up and took him to a rehabilitation center.

"Our goal was to release him back into the wild when he was ready, in the exact same spot where he had been found."

Yumbo remained a prisoner for two years, and then he did indeed return home to Yunguilla. The local media was there in 2015 when Yumbo was brought back to the Andean village in a transport crate in the back of a pickup truck. The now-grown bear disappeared into the forest.

"Yumbo is the first bear in Ecuador to be successfully reintroduced to the wild," Santiago told me proudly. "The other bears around Yunguilla accepted the newcomer. They probably smelled that he was related to them and said, 'Hey, we know you. Welcome back!'"

SINCE THEN, ABOUT five thousand tourists a year come to the bear village of Yunguilla. The majority of them, mostly Ecuadorians, make it a short side trip, but lots of foreigners book a number of nights in one of the beautiful guest accommodations. The village restaurant serves biodynamic products grown by local farmers. Many visitors also buy local delicacies. The local cheese—*queso de Yunguilla*—and tropical fruit jam are popular items for visitors to take home with them when they travel back to Quito. The inhabitants of this little Andean settlement live well off this "authentic village experience," as they call their form of tourism.

The slopes around Yunguilla are greening up again, while elsewhere in the Andes the mountain forests are slowly being cut down. Hardly anyone here cuts down trees. In fact, the villagers in Yunguilla are helping the forest regrow. The village has its own tree nursery, where the locals grow seedlings of native trees, which they then plant on the mountain slopes. For a long time, biologists thought it was impossible to bring the forest back. But it can be done if people help the process along. It is not enough to wait for the trees to grow back over the cow pastures by themselves. The first generation of trees must be laboriously planted. Then seeds brought in by birds and bats can sprout in their shade. The villagers of Yunguilla have reforested hundreds of acres to date. Just as the people have benefited from the bears, now the furry animals are in turn benefiting from the people. It's a win-win playing out right next to the capital, Quito.

It's important to remember that the spectacled bears are not the only animals to profit from the symbiotic relationship between humans and bears. Where forests are returning around Yunguilla, they provide habitat for pumas, wild pigs, hummingbirds and other birds, frogs and turtles, bats, and Andean tapirs, as well as thousands of species of insects.

Yunguilla's successes demonstrate that if you want to save individual species like Andean bears, you get the most benefit if you keep the whole ecosystem intact—and that conservation often also makes economic sense for the local people.

Santiago Molina is thinking one step ahead. His dream is that other communities will follow Yunguilla's example and get involved with reforestation projects. If they did, then what are now isolated patches of forest would grow together to become, once again, a huge paradise for bears, the *corredor ecológico del oso andino*—or ecological corridor for Andean bears.

This kind of reimagining of the relationship between people and animals is not something that can happen without the occasional setback. Every once in a while, a cow falls victim to a bear. Even the high-profile repatriated bear Yumbo is known to have helped himself to dinner in one of the cow pastures. When something like that happens, a few of the Yunguilla farmers wonder if the path they have chosen is the right one. But then all they have to do is to compare their living conditions to those of their parents and most of their doubts vanish.

IN THE IMPOVERISHED world of Ecuador's mountains, small wonders like this are happening in other places as well. The brothers Angel and Rodrigo Paz were also once farmers. Like most who make their living this way, they lived off about ten dollars a day, which they earned by selling milk, eggs, and vegetables. Their land, not far from the little town of Mindo, also had a small piece of forest on it, which at certain times of the year was visited every morning by a flock of birds that put on quite a show. They were Andean cocks-of-the-rock meeting up in what is known as a lek, a display ground where the males gather to show off to potential mates.

During mating season, male Andean cocks-of-the-rock make a great deal of noise every morning and evening as they jump on and off branches and take turns showing off their brilliant red breasts and head feathers to each other. Females are attracted to the spectacle and size up the puffed-up cocks. When the males notice a female showing some interest, they amp up their already wild displays.

For a long time, the birds' wild mating rituals and showy plumage were a knotty problem for the evolutionary biologist Charles Darwin. They just didn't seem to fit in with the theory of the origin of species that he had so painstakingly developed.

Birds like cocks-of-the-rock seem to burn up valuable energy in their daily competitions to no discernible end. Not only that, but their displays put them in danger. Attracted by the birds' shrieks and colorful feathers, hawks, pumas, ocelots, and snakes such as boa constrictors constantly help themselves to birds on the display grounds. And yet the spectacle is worthwhile for the cocks that stand out from the others. The females select the most opulently colored super-dancers among the males and allow them to mate with them. The higher success with the females is their compensation for the energy they invest in wild behavior on the display grounds.

Eventually, Darwin recognized this and described sexual selection as one of the forces driving the evolution of species. This force of nature causes crickets to burst into song, deer to grow huge racks of antlers, and peacocks to unfold their shimmering, colorful tails—and Andean cocks-of-the rock to dance wildly on their communal display grounds.

~~~~~~~

THE PAZ BROTHERS, Angel and Rodrigo, wondered whether the morning show on their farm might attract a very special kind of tourist—birders. In 2005, they cleared a path through their forest and established the *Refugio Paz de las Aves*, complete with a small guesthouse for birders. A month passed and the path grew over without a single tourist having stumbled across Angel and Rodrigo's—and their wives Maria and Diana's—new venture.

Things would soon change. Gradually word got around that if you visited the Paz family's refuge, you would almost certainly see the big show put on by the *gallos de la peña*, the Andean cocks-of-the-rock. But it was another bird, the giant antpitta, that made Angel Paz a true legend among birders.

In 2018, I booked myself in at the *Refugio Paz de las Aves*. The things I had heard about the antpitta show at the Paz brothers' refuge were just amazing and I simply had to see for myself. We set out early in the morning. Angel and Rodrigo led me and a group of U.S. tourists into their forest. We stopped in a clearing. Angel spread a handful of mealworms along a fallen tree trunk, pursed his lips, and whistled his mating call. That brought Maria out into the open. A bird that looked like an oval football on stilts left the protection of the undergrowth and ran on long legs directly to the mealworm station. The giant antpitta is, in comparison with the magnificent Andean cock-of-the-rock, a rather drab bird. Still, the U.S. birders excitedly brought out their cameras with telephoto lenses attached and photographed the bird like a pack of excited children. All around me, digital cameras clicked and beeped.

It is thought that only 2,500 giant antpittas still stride through the mountain forests of Colombia and Ecuador. Angel Paz just happened to come across one of them while he was walking through his forest one day. He called the bird Maria and decided that he and the little lady should become friends. On numerous occasions, he brought squirming treats for Maria into the forest, until the bird gradually overcame her shyness and reciprocated Angel's affections. Today, Angel Paz and the antpitta Maria are stars and an important stopping point for wealthy bird-watchers traveling in Ecuador. Lots of people book the antpitta tour and make the stop at the *Refugio Paz*.

For a long time, people thought the giant antpitta no longer existed in large parts of Ecuador. On our tour, after our encounter with Maria, we saw other giant antpittas, close relatives such as the bicolored antpitta, many different species of hummingbirds, and—of course—displaying Andean cocks-of-the-rock, much to the delight of the group of visitors from the United States. Maria and the other birds haven't made the brothers Paz and their families rich, but they have assured them of a good income. And, just like in Yunguilla, the forest around the *Refugio Paz de las Aves* is growing rather than shrinking—even though the refuge is not a state-sponsored conservation area. The forest is returning, solely thanks to the giant antpitta Maria and Angel and Rodrigo Paz's initiative and love of nature.

Places such as Yunguilla and the *Refugio Paz* are still the exceptions in Ecuador. They do, however, demonstrate how much change the initiatives of individuals can bring about. On the western slopes of the Andes, there are no government-sponsored national parks, which makes private conservation areas all the more important.

~~~~~~

THE GIANT ANTPITTA also tiptoes through the *Reserva Otonga*, the forest conserved by the Italian biologist Giovanni Onore where I once spent an afternoon battling a great swarm of army ants with a broom. The reserve is a very special conservation area with a unique story.

Until his retirement, Giovanni Onore was a professor of biology at the Pontifical Catholic University of Ecuador in Quito. While he was still working, he was shocked by how quickly the rainforests in the small country were being cut down.

"In the 1980s when I first visited the area around the Toachi valley, where the *reserva* is today, it was one great expanse of paradise. Forests full of orchids and birds were everywhere."

Onore, still spry at eighty, told me this as he was trudging up a path in his cloud forest at an insanely fast pace.

"Poverty forced the people who live here to cut down the trees to create more pastures for cows. Back then, I believed nothing could be done to stop it."

But the Italian, as it turns out, was wrong. He was lucky to find a kindred spirit in the Toachi valley. César Tapia was a cattle farmer, like almost everyone else in the region. But he too harbored a love for nature, plants, and the many animals in the forest. It made him unhappy to see it all disappearing. The two men were friends for almost forty years and they made a pact. The biologist supported the farmer and his family and made it possible for César Tapia to send his nine children to school. In return, César promised to help Giovanni. Together, they wanted to save what was left of the forest.

Onore gathered donations from his native country of Italy, and César connected the professor to other farmers in the community. Although many farmers viewed conservation as a strange obsession of rich countries, they were more than happy to do business with the Italian. With César's help, Giovanni bought up unused plots of land. He found remnants of cloud forests, especially in places where the land was particularly steep. Gradually, a unique reserve came into being—the *Reserva Otonga*.

In contrast to Yunguilla or the *Refugio Paz*, the *Reserva Otonga* was intended to be not a forest for tourists but a forest for scientists. Giovanni invited researchers from Ecuador and around the world to work in his conservation area. Usually, César or his sons guided guest scientists through "their" forest. They helped biologists from other parts of the world discover and describe hundreds of new species of plants, spiders, and insects. Then, a mammal—an animal that looked like a teddy bear with its furry coat and button eyes—brought international attention to the Otonga rainforest.

~~~~~~~

THE STORY STARTS in the United States. Biologist Kristofer Helgen and his team from the Smithsonian Institution's National Museum of Natural History in Washington, D.C., wanted to do what is known as a "revision" of the descriptions of a group of mammals called olingos. There were three known species of these omnivores, which live in South and Central America and are related to raccoons. They are called olingos because of their strong odor. The name is derived from the Spanish word *oliente*, which means "smelly."

Olingos are blessed with a long tail, one of many adaptations to life in the canopy of tropical forests. They use their tails for balance when they travel through the trees, looking for food in the form of fruit or small animals. As all species are nocturnal, they'd generally attracted little attention from tropical biologists. That is, until Helgen and his olingo revision.

In a revision, biologists examine the museum specimens used to describe different species within a genus. Sometimes a species is discovered more than once and therefore incorrectly assigned two different species names. Conversely, animals assumed to belong to a single species may actually represent two separate species. For a long time, whether the museum specimens were considered to be variants and subspecies or different species was a matter of weighing the evidence at hand, and, for many animals, the answers to these questions are not always obvious.

In the course of their research, Helgen and his team came across a series of specimens that were different from other olingos. They were smaller and more compact; their fur was denser and their heads weren't quite as large. Was a hitherto-overlooked fourth species of these smelly furballs hiding in plain sight?

However, before the team became the first zoologists in thirty-five years to describe a new mammal species in the order Carnivora, they had to be absolutely certain. These days, in cases like this, biologists can turn to genetic analysis, which makes the decision to name a new species somewhat more objective than simply measuring jawbones or counting bristles and feathers.

After the genetic analysis, Helgen was certain. He had discovered a fourth species of olingo—in a museum. The researcher named the fourth olingo *Bassaricyon neblina*, but it became known by its common name—olinguito, or "little stinker."

Helgen did some research to find out where the museum specimens originated. Almost all had been stored for more than a century and many had been captured in the northern Andes—in the cloud forests of Colombia and Ecuador at an elevation of about 6,500 feet (2,000 meters). The craziest thing was that at least one female olinguito had lived in zoos in the United States in the 1960s and 1970s. It was a case of mistaken identity. Everyone assumed she was one of the known olingos, and so zookeepers attempted to mate her with olingos of the wrong species—an endeavor that, of course, had no hope of success.

KRISTOFER HELGEN WANTED to find out if there were still olinguitos living in the wild. Chance came to the rescue, as it so often does in life. The Ecuadorian biologist Miguel Pinto was doing a work placement at the world-renowned Smithsonian National Museum of Natural History. Pinto was a student of the biology professor and rainforest savior Giovanni Onore and just happened to be in Washington, where he was continuing his training. Helgen assigned the intern to the olinguito expedition in Ecuador.

Perhaps you're thinking Miguel Pinto's expedition in search of the "little stinker" cost thousands of dollars and had umpteen participants, vehicles, and porters. It did not. He traveled alone through Ecuador and his equipment consisted of a powerful flashlight and a video camera. Naturally, one of the places he searched for the olinguito was the cloud forest in the *Reserva Otonga*.

Pinto prepared himself for a long, exhausting expedition that he expected might last many nights. He set off with César Tapia as his guide. On his first night out, after just a short hike, he spotted a small creature hopping from branch to branch. He managed to capture a few seconds on video and the footage confirmed that he had found what he had been sent to search for.

And with that, everything fell in place. The research team invited journalists to press conferences in Washington, D.C., and Quito to announce the sensational news that a new species in the order Carnivora had been discovered and described for science—even if it was just a small nocturnal stinker, the olinguito.[35]

For the biologist Giovanni Onore, every new species discovered in the reserve is of equal importance, whether it be a mammal or an insect. On the one hand, the protected area offers sanctuary to many animals and plants; on the other, every newly discovered species helps secure the future of the reserve. That is exactly Giovanni's strategy.

"That's why I invite researchers to visit," Giovanni told me. "If they find something new and call the species *otongico* or *otongensis*, after the reserve, then it's clear that the area should be left untouched because of its biodiversity. That's how I protect it. Scientists succeed where laws and politicians fail."

The cloud forest in Otonga, however, is not only beneficial for bears, antpittas, olinguitos, and thousands of species of insects; it is also beneficial for the people who live near it. It was important to Giovanni Onore that the farmers in the region also profit from his project. After all, they had to get accustomed to some restrictions on their lifestyles. They had been used to going out into the forest to cut wood and to hunt birds, armadillos, deer, and other wild animals. That was all suddenly prohibited. Onore looked for other ways to get the local inhabitants on his side and excited about nature conservation. Like the brothers Paz, he settled on paying visitors. Giovanni's Otonga Foundation built a smart little jungle hotel for scientists and students. The biology professor employed only local people to build it. The scientists' research area—a beautiful cloud forest full of animals and plants that have barely been studied—lies right outside the front door.

Onore is optimistic about the future.

"The hotel and guests will generate enough income to keep the forest going. We need people to patrol the forest, people to keep the paths clear, people to check that no one cuts the forest down, and people to welcome the guests. That means employment and also income to fund the forest."

There's more. The trees in the forest gather water from the clouds and clean it. Local politicians have recognized the "ecosystem services" the region provides.

"Nature is valuable. More and more people are coming to understand that. Officials from local towns are coming here to look for water for their residents," Onore told me with pride. "And doctors are saying, 'Let's protect the forests around the places where we build hospitals.' Clean water guarantees fewer stomach and gut disorders. There's a connection between people and nature conservation."

~~~~~~

IF YOU FOLLOW these thoughts to their logical conclusion, you could also say that a small Andean olinguito, a little stinker, can help save people—by ensuring that the cloud forest remains intact.

It bears repeating: if we want to protect rare animals and plants, we need to protect their habitat. The examples from Ecuador are a testament to this. And what is possible in South America, which is not exactly blessed with wealth, should also be possible in other parts of the world, right?

21

A Network of Habitats, Large and Small

Conservation doesn't just happen in exotic places far from home. What people do in the regions where they live—out in the countryside, in cities, and in their own backyards—is also important. These efforts can involve individuals, organizations, and civic authorities. They can range from conservation to restoration, from pockets of diversity to projects on a grand scale.

~~~~~~

SOME IDEAS ARE simply so good that you wonder why no one thought of them before. In 2019, a German farmer in Lower Saxony called Karsten Padeken had one of these inspired moments.

I've already touched on how detrimental modern agricultural practices are to nature. It is easy to point the finger at farmers: they are the ones killing bees, poisoning groundwater, and causing the once-rich diversity of species to disappear.

Intensive agriculture is, of course, a problem for nature, but it is not helpful to make farmers take all the blame. Indeed, laying blame is easy. It's more productive to try to find ways out of the mess we are in. Karsten Padeken, the farmer from northern Germany, is doing just that.

Padeken is not an organic farmer; he farms his land using conventional methods. When he read about birds and insects dying, however, he did not dismiss the problem and think it had nothing to do with him.

"I realize," he told me, "that there are problems. We all need to pull together to do something about them." I asked him who he meant by "we."

"Well, all of us. You can't just say it's the farmers who are responsible." People have to get their food from somewhere and farmers need to pay their bills. "If I buy land and pay off my loans, the last thing the bank cares about is what I'm doing for insects."

Calculating income was just part of the equation for Karsten Padeken, however, and he decided he was going to do something. He set aside three-quarters of an acre (three thousand square meters) of his land for insects and birds. Instead of planting maize or potatoes, he planted wildflowers. But he also wanted other people who lived in the area to get on board. He made them an offer. For just 25 euros (about U.S. $30), anyone can sponsor one of a total of 150 strips of flowery meadow.

"If you don't count the extra work I put into this, I just about manage to cover the income I lose by not harvesting anything from this land. But only if I sell all 150 shares."

But, for Padeken, it's not all about the money. It's important to him that people do more than simply point their fingers at farmers. Making fine speeches is all well and good, but if you're really going to take responsibility, you need to step up and do something.

Just one week after the local newspaper first reported on his project, 40 of the 150 shares had been sold. I wish him success and hope other people might try something similar. And wouldn't a couple of strips of wildflowers make a more beautiful present than a few roses flown in from Kenya or Colombia that are going to wilt in the vase after a few days? Of course, three-quarters of an acre (three thousand square meters) of wildflowers in northern Germany are not going to be enough to save rare native bees or stop birds like the lapwing from disappearing. But Karsten Padeken's action is a start and, ideally, one of many.

The good news about the loss of species is that everyone can do something to help preserve at least a small piece of the biodiversity around them. And the more people who do this, the better. Now is a good time to start. Many people are learning to appreciate the insects they used to love to hate. Well, maybe not cockroaches, but at least bees, butterflies, and ladybugs.

~~~~~~

YOU CANNOT PROTECT animals unless you also protect the places where they live. For many insects and birds, the area does not necessarily have to be large. You could have a meadow with wildflowers or scattered fruit trees, a small pond, or woodland edges and clearings where fallen trees are left to decompose. The more mini-habitats like this are not only left but also created, the denser the network of islands of biodiversity across a country—and the more both plants and animals benefit. Protecting biodiversity in our manipulated landscapes really can be a grassroots revolution. The more people who participate—from farmers to gardeners, schools, cities, communities, and counties up to politicians—the better. Everyone has a chance to be an eco-warrior and conservationist.

A first step is to re-vision how we want our urban landscapes to look. It's up to you to decide. Do you want a neat and tidy garden with grass cropped as short as Bruce Willis's hair? Or do you want a mini-wilderness with wildflowers, native shrubs, and corners where even plants like stinging nettles—supposedly the nemesis of home gardeners—also have a right to exist?

On the other side of the Atlantic from Padeken's plots, there are similar problems—and an increasing number of gardeners who are interested in offering butterflies, native bees, and many other rare insects a safe haven and an ample supply of food in their backyards. Over ten thousand gardeners in both Canada and the U.S. have answered the call of the Xerces Society and made their property insect-friendly by planting flowers and shrubs that provide pollinators with nectar and caterpillars with food. The Xerces Society was founded in 1971 with the goal of protecting invertebrates. The society hopes to save other insects from going the way of the butterfly from which the organization takes its name. The Xerces blue was found among the coastal sand dunes of Northern California until the early twentieth century. Its habitat, however, was increasingly overtaken by wealthy neighborhoods. By the 1940s, the sand dunes had almost completely disappeared—and with them the Xerces butterfly, the first butterfly known to have gone extinct in the U.S. as a direct result of human activity.

The Xerces Society advises home gardeners on how they can manage their property without using pesticides, how to provide nesting sites for insects, and which plants to grow—whether they are gardening in Canada, California, the Gulf Coast, or anywhere else in North America. The environmental organization gives tips to all who are interested and recommends plants and seeds suited for every growing zone. They do all this in the hopes that as few insects as possible will share the Xerces blue's sad fate.

~~~~~~

BUT HELPING INSECTS is not the job of home gardeners alone. Towns and villages should also be asking themselves whether it still makes sense for public spaces to emulate British parks of past centuries. A few years ago, when my hometown of Cologne began to plant a few green strips along Innere Kanalstraße (Inner Canal Street), a busy street that circles downtown, many residents of the cathedral city rubbed their eyes in amazement. It didn't take much to bring a bit of nature right into the middle of town. "Urban meadows, not lawns," is the mantra for one of the projects spearheaded by NABU, the regional conservation organization, in partnership with Cologne's department of landscape maintenance and green spaces. Rare plants such as harebells, meadow campion, cornflowers, bright sky-blue native meadow sage, and cowslips are once again at home in the middle of the city. And with them, it is hoped, will come rare butterflies like hairstreaks and day-flying burnet moths— also known as "blood drops" in German because of the striking red spots on their wings. As these natural-style meadows need mowing only once or twice a year, they are not more expensive for the municipalities that plant them.

London, the venerable capital of Great Britain, has gone much further than my hometown. When the city was preparing for the 2012 Olympic Games, the authorities decided nature would also profit from the tremendous changes the metropolis was facing. An abandoned industrial site in the valley of the river Lea, a tributary to the Thames, was converted to Queen Elizabeth Olympic Park, which was to be the center of the Olympic complex. Not only was the huge Olympic stadium built there, but also smaller venues for basketball and water polo. Nature-loving Londoners were lucky that the landscape designers responsible for the park, Nigel Dunnett and James Hitchmough, two professors at the University of Sheffield, had thought about what would happen to the area once the mega-event was over.

From the outset, the idea was that some of the sporting venues would be disassembled. Wildflowers were to bloom in the Lea valley—and it was to accommodate four thousand trees and expanses of wetland. The park was designed not only as a space for plants and animals but also as a shining example of what is possible in a city. The planners hoped that people who strolled through the park every day would take the idea of natural landscape back to their own gardens. The park, it was hoped, would be a catalyst for a much larger movement. "There are 13,000 square kilometers of gardens in the U.K., more than all the official nature reserves put together—so gardens are our nature reserves," said Professor Dunnett at the beginning of the project. Unfortunately, no one has counted how many British gardeners have been inspired by the park to transform their own properties. Nevertheless, the park itself is flourishing.

In 2020, just twelve years after the onset of the project, came the first signs that it was indeed a success. Six of Great Britain's most protected birds were seen in the park. It's extraordinary to think that rare birds such as kingfishers, black redstarts, Cetti's warblers, fieldfares, redwings, and peregrine falcons have all been spotted in the center of London. Where once athletes sparred in Olympic competition, 1,100 species of invertebrates (91 of which are protected) now scramble, crawl, and fly. There is even one species of beetle reproducing in the park that was thought to have died out in the U.K. The streaked bombardier beetle is not only a rare insect but also one with an astonishing skill. When it feels threatened, it releases a volatile mix of gases. When the beetle's stink bomb explodes, the sound is audible even to the human ear. The combination of gases and the heat of the blast (over 212 degrees Fahrenheit or 100 degrees Celsius) cause almost all potential predators to turn tail and run. But that's not the only time the beetles play with fire. They also let their gases rip to attract partners when they are in the mood to mate. Undisturbed areas all over the Queen Elizabeth Olympic Park are now among the few places in Great Britain where the little beetles crawl around letting off their small explosions.

Insect habitat can be planted whenever there's a will to do so—even in the most densely developed places in the world. Raised on steel supports thirty feet (over nine meters) above street level, the High Line stretches like a dragon's tail for almost a mile and a half (2.33 kilometers) through New York's Meatpacking District on the western side of the island of Manhattan. Freight trains once traveled the High Line, but the line was shut down in 1980. In 2006, work started to create the narrowest public park in the city. The final section was completed thirteen years later.

In 2017, entomologists Sarah Kornbluth and Corey Smith from the American Museum of Natural History took the first inventory to document whether insects were visiting the still-new park in the middle of an ocean of buildings. Even though the park had not been there long, the two biologists counted thirty-three species of native bees in the small strip of green. More would join them. Kornbluth and Smith advised the landscaping team to build more nesting sites—so-called insect hotels—in the park. New Yorkers and bees alike benefit from this elevated greenway. The bees find nectar and places to nest and the human visitors can enjoy a strip of nature in a desert of concrete—even if the strip is very, very narrow.

To effect the kinds of changes being seen in Cologne, London, and New York, all you have to do is bid adieu to monotonous ornamental plantings—at least in some parts of urban parks—and tolerate a little chaos and wildness. I couldn't agree more with this statement by Cologne's NABU, the activists for more wilderness in the parks of my hometown:

> Meadows full of wildflowers have a value that is not to be measured in dollars and cents. And for that very reason, we need them more than ever.

~~~~~~

EVEN WHEN GARDENS and city parks are converted into small pieces of wilderness, that alone is not enough to halt insect—and bird—decline. It is just as important to protect the natural areas that remain in the landscapes we have shaped. Thankfully, there are people who remind us how important that is and who are ready to put themselves at the forefront of the fight for wilderness areas.

In the United States, Joan Maloof is one of these warriors for wilderness. For half her working life, she was a professor of biology at Salisbury University in Maryland, where she researched and taught about the value of what are known as old-growth forests. When people think of old-growth forests, the first images that come to mind are usually of vast undisturbed tropical forests in the Amazon, Central Africa, or Southeast Asia. What is less well known is that even in an industrialized country like the U.S.—where the stories we usually hear are of the settlement of the continent by Europeans, of clear-cutting forests, and of nature transformed—there are still forests, even outside national parks, where trees have not been felled for centuries.

Today almost all the old-growth forests in the U.S. are gone and instead of being shaped by forests, the land is now shaped by towns, roads, fields, and ranches. But forests have not completely disappeared. Remnant patches of old growth remain—although you almost need a magnifying glass to find them on a map. When I interviewed Joan Maloof, she told me that 99 percent of the original forest in the eastern states no longer exists. Things are only marginally better in the West, where only 5 percent of the old-growth forests have not yet fallen victim to chain saws.

"A forest is a forest," people kept telling Maloof every time she wanted to champion a forest and stop people from cutting down the trees and "managing" the forest as though its only reason for existence was to produce timber. "One big problem is how students are trained in forestry schools," Maloof told me. Economic concerns loom large. There is far less discussion about the ecological value of a forest. And old-growth forests have enormous ecological value. The big difference between an old-growth forest and a managed forest is that trees in an old-growth forest are left to grow until they die a natural death. Ancient, rotting trees are a paradise for countless numbers of insect larvae. Birds find food and shelter there. Some birds that are now rare, like the northern spotted owl, for instance, are so sensitive to human disturbance that they nest and hunt exclusively in old-growth forests.

Other animals are becoming increasingly scarce as more people encroach on their habitat. Red-backed salamanders prefer to crawl through the leaf litter and woody debris on the floor of "unmanaged" old-growth forests in the eastern U.S. Where foresters don't go through the forest "tidying up," you can sometimes find as many as five hundred of these salamanders per acre. In a second-growth forest close by, in contrast, a similar area might offer protection to only one hundred individuals. If the forest is cut down, the amphibians have nowhere else to go and disappear forever.

Over the course of her university career, it became increasingly clear to Joan Maloof that research alone was not going to be enough to save the last old-growth forests from loggers. But whenever she wanted to step up to protect a piece of forest, she found she didn't have the time. There were students to supervise and scientific papers to write. Finally, Maloof made a radical decision. She retired early in 2011 to start her own conservation organization: the Old-Growth Forest Network. Her goal is to identify as many old-growth forests as possible in the United States and then make sure they are protected.

"You can't always just point a finger at countries like Brazil," Maloof explained. "We need to get involved here too, precisely because there are so few primary forests left here." Her goal is to have at least one protected forest in each of the counties in the U.S. where forests can grow (about 2,300 out of the total 3,140 counties) be part of the network. Maloof and the four people she now employs travel tirelessly around the country giving presentations. And after every presentation, more people step up to volunteer or donate money to contribute to the project's success. One hundred and sixteen counties in twenty-four states have already signed up.

"Even the smallest forest that is preserved can provide an important instigator of change," Maloof told me. These islands of forest not only provide refuges for rare plants and animals. From them, their inhabitants can spread out and conquer new habitats, if other forests in the area are also allowed to age naturally.

"We are not against logging," Maloof was careful to explain. "We just don't think it should be happening everywhere." When she has an opportunity to walk through one of the old-growth forests she has helped protect, she is reminded that she is doing the right thing. The sight of an ovenbird nest with four tiny eggs makes her heart swell. She knows that many migratory birds and their offspring will come back to this small patch of forest because they know that in this place, *Homo sapiens* has left their world intact.

~~~~~~~

IT'S IMPORTANT TO protect what wilderness remains no matter where you live. Many conservationists, however, go one step further. They want to restore what has been destroyed over the centuries. A particularly spectacular example is the return of a vast part of the American landscape: prairie grasslands.

As far back as the nineteenth century, people were trying to conserve the once seemingly endless grasslands called the Great Plains for future generations. The prairies lie in the rain shadow of the Rocky Mountains. For thousands of years, sparse rainfall and millions of grazing mammals, like bison and pronghorn, made it almost impossible for trees to grow there—which opened up space for 1,500 other plant species to grow instead. Wolves, bears, and coyotes roamed the grasslands and millions of prairie dogs dug their burrows in the wide-open plains. The prairies ensured the survival of 350 species of birds and 220 species of butterflies.

But then the bison hunters and farmers arrived. And the native prairie grasslands shrank. In 1984, only about eight hundred bison still lived in all of North America, and two hundred of them were in Yellowstone National Park, the last free-roaming bison in the United States. Their numbers fell to an all-time low in 1902: only twenty-three animals remained. The grasslands had been transformed into farmland—the amazing prairie, a symbol of America, seemed to have been lost forever.

Conservationists, including many Indigenous tribes, decided to fight back. In the 1990s, fifteen new bison herds were established, most of them on Indian reservations. Today, there are 350,000 of these powerful grazers. At the beginning of the twenty-first century, the largest wildlife preserve in the United States was established in Montana. Covering three million acres, the American Prairie Reserve is as large as Connecticut. The private American Prairie Foundation receives donations from across the country. These donations allow them to purchase targeted parcels of land, including entire ranches, from private landowners, so corridors can be created to connect what were once isolated public nature reserves. Fences that stop wild animals from roaming freely are removed, and in 2005, the first bison were released into the wild.

The American Prairie Reserve is not just a project to protect bison. The much larger goal is to restore a complete, self-sustaining ecosystem. And the chances are good that the project will succeed. Not only bison, pronghorn, prairie dogs, swift foxes, and mountain lions will benefit, but also millions of butterflies, grasshoppers, spiders, and other invertebrates. One can never say this too often: you can only successfully protect plants and animals if you also protect their habitat and—if necessary—restore it.

~~~~~~

THE AMERICAN PRAIRIE Reserve may be immense, but it is worth thinking on an even larger scale. Animals do not recognize borders and, for them, human constructions such as nature preserves and national parks simply do not exist. Migratory birds fly for thousands of miles; wolves can travel more than sixty miles (about one hundred kilometers) in a single night. The smaller and more isolated a nature preserve is, the less it is worth. But, of course, we can never turn the clock back to the time when North America and Europe were uninterrupted wilderness. And therefore, it is all the more important that we connect existing conservation areas in clever ways. Forests, hedges, wetlands, riverbanks, and grasslands can then become corridors along which habitats can be exchanged and enriched.

Around the world, there are increasing numbers of corridors that connect areas of biological diversity. I've already mentioned Ecuador's corridor for Andean bears. The joint American and Canadian Yellowstone to Yukon Conservation Initiative is an example of a particularly ambitious dream. What if federal governments, local authorities, conservation organizations, communities, and private landholders all came together to connect the impressive Yellowstone National Park with the equally impressive Yukon Territory in northern Canada in such a way that wild animals encountered as few obstacles as possible when traveling from one to the other by way of the Rocky Mountains—a distance of nearly 2,000 miles (over 3,200 kilometers)? It sounds crazy, but 450 partners on both sides of the border have been working to make this dream a reality since 1993. New wildlife corridors now connect existing national, state, and provincial parks; degraded forests are being restored; wildlife crossings are being built across busy highways; and—one of the most important initiatives of the project—efforts are being made to get local residents behind the idea of wilderness that stretches free from any human-imposed impediments. Free passage for wild animals from Yellowstone to the Yukon, this is how I see the modern conservation movement developing.

Protecting, restoring, and connecting habitats seems to me the only way we can stop the catastrophic mass extinction of species on our planet.

22

Unexpected Events in a Gravel Pit

n all of this, we should never underestimate the resilience of nature itself. Occasionally, nature gets a second chance where you least expect it, sometimes with no help from us at all. I've already mentioned the Upper Rhine Valley and how radically people have altered the landscape.

The river, which once wove its way through the landscape, has been forced into a single channel. Old side channels have dried out—and over time, an enormous wetland area has been transformed into agricultural fields and much biodiversity has been lost. But that's not all. In an assault that continues to this day, enormous excavators dig deep wounds in the ground as they remove sand and gravel to be used as construction material. As the water table along the river is high, these gravel pits immediately flood. You wouldn't expect to find an animal or plant paradise in such artificial bodies of water, but every once in a while, you are surprised...

IN 2014, I met the biologist Serge Dumont. A professor of biology at the University of Strasbourg, he is also a fantastic diver and an extraordinary photographer who specializes in underwater films. Serge Dumont was going to open my eyes a little further when it came to viewing nature. It's not just species that can be small and overlooked; sometimes whole habitats are underestimated. Their beauty is obvious only to those who take the time to scrutinize them more closely. And that is exactly what Serge Dumont does.

Thomas Weidenbach, a film producer from Cologne, introduced us for an unusual project: a nature documentary about life in a flooded gravel pit. The first time we met, Dumont explained to me that of course he also loved to dive on tropical reefs full of colorful fish, but right now he was living in the Upper Rhine Valley and not in Indonesia, Costa Rica, or Australia.

I wanted to know if an artificial body of water would provide enough material for a forty-five-minute nature film.

"Most flooded gravel pits aren't anything special," he said, confirming my worry. "Where people swim in them and keep trampling down the vegetation on the banks and where anglers stock them with massive amounts of carp, they do look bad."

Luckily, however, he knew of a couple of pits that were off the beaten track and therefore spared by anglers and hordes of swimmers. Nature had been able to reclaim them.

The diving professor invited me back to his home and showed me his best photographs and footage. And I couldn't believe all the things you can find in a supposedly lifeless gravel pit.

~~~~~~

SERGE DUMONT IS a researcher behind a camera. For him, each subject is equally beautiful, whether it be an impressively large wels catfish or a freshwater jellyfish half the size of your little finger. He films everything from endangered freshwater eels to tiny bryozoans that use even tinier cilia to maneuver the suspended particles they eat into their mouths. Dumont sometimes dives right to the bottom of the ponds, which can be up to 165 feet (50 meters) deep. There he documents the mysterious world of decomposers: animals that wait for organic debris to sink down into the darkness of their world.

Flatworms, red sludge worms, and tiny freshwater crustaceans feed off dead plants and animals. Freshwater sponges thrive in gravel pits and cover the bottom like colorful pebbles. It was clear no one had ever shone a spotlight on these seemingly insignificant spaces in quite the way Dumont was doing.

Dumont uses a rebreather when he dives—an extremely quiet breathing device that produces no bubbles. He doesn't pursue the animals. Instead, he waits. He remains motionless—often for minutes—until the fish and waterfowl swim to him, which allows him to capture behaviors never seen before.

And that was how a pair of great crested grebes, waterbirds with lobed toes, became the unexpected stars of our film. These birds, about the size of a duck, hunt underwater, mostly for small fish. That much was already well known. But no one had ever observed how these birds manage to snap up agile fish adapted over millennia to life underwater. Great crested grebes are cautious birds. Before they dive, they stick their heads underwater to see if there are dangers lurking below. If they catch sight of something unusual—a filmmaker by the name of Serge Dumont, for example—they keep on swimming and choose a different spot for their underwater chase.

But Dumont didn't give up. He kept putting on his rebreather and diving down into the sleek waterbirds' hunting territory. Over time, the grebes began to accept the presence of their unusual guest and eventually hunted directly in front of his camera. Dumont revealed for the first time how these birds manage to snatch their prey. The underwater hunters search for fish among aquatic plants, in exactly those places where fish try to hide. The fish are cornered and their hiding place becomes a deadly trap. And to Dumont's surprise, the birds swallow their prey while still underwater and keep hunting until they run out of air. Nature can still surprise us—even on our doorstep in an environment as stark as a gravel pit.

In our nature documentary, Serge Dumont drew back the curtain on a magical world—a sanctuary for fish, birds, crustaceans, and insects right in the middle of the ravaged Upper Rhine Valley. Of course, a few gravel pits cannot replace what it has taken humans centuries to destroy along the length of the Rhine. But recognizing the ecological value of such wetland areas and preserving them is at least a start.

As I mentioned already, the water quality of the Rhine has improved considerably in the past few decades thanks to modern sewage treatment plants. And in many places, the river is once again being allowed to revert to its old meandering ways. Dikes are being breached, and side channels cut off for decades are now once again connected to the main river. This creates new habitats full of hiding spots and places where fish can lay their eggs.

More than fifty species of fish, almost as many as once swam here, can now live once again along the length of the Rhine. Riparian woodlands that flood every time the river reaches its peak are thriving once more.

The Rhine's comeback is still in its early stages. There are still far too few healthy wetlands along the upper reaches of the river. Moreover, overfertilization in the surrounding agricultural fields leads to excess nutrients leaching into groundwater. These nutrients end up in sensitive areas, where they promote the growth of algae. The algae then proliferate and overrun everything, and their rampant growth is now the most pressing problem for biodiversity on both sides of the great river.

SERGE DUMONT STARTED as a professor of biology. Over the years, he has become a nature documentary filmmaker and an environmental activist. We have now made two films together and he has traveled around Germany making public presentations to tell divers, anglers, farmers, and community leaders about the treasures out there, treasures that are worthy of their protection.

Dumont has observed pikes mating and trained his lens on frog-eating water snakes and the last western curlews in Alsace. I have to say—and how could it be otherwise—that my personal favorite in Dumont's underwater cosmos is a somewhat nondescript insect: the common blue damselfly. Damselflies are somewhat smaller, more delicate versions of dragonflies. One easy way to tell them apart is that dragonflies at rest leave their wings extended, whereas damselflies fold them together.

When biologists call something common, they are not making a comment about its behavior or even its character. In biologist-speak, it merely means that it is commonly found and widely distributed. The common blue damselfly—finally some good news—is still common in central Europe. Its behavior, however, is far from common. Indeed, it is most unusual.

On one of his many dives, Dumont spied an adult female common blue damselfly in the middle of some aquatic plants in over three feet (one meter) of water. It is well known that the voracious larvae of all damselflies (and the larger, related dragonflies) live underwater. But to find a diving adult surprised even the biologist Dumont. He decided to take a closer look at the damselfly's puzzling behavior—with his camera, of course.

Almost all damselflies and dragonflies mate close to the body of water where their larvae will grow up. After the dragonfly wedding, the females look for a place to lay their eggs. Some simply drop their eggs while flying over the water. Others land on aquatic plants, dip their slender abdomens beneath the surface, and bore a hole in a plant stem, where they lay their eggs. In many dragonfly species, the male doesn't leave the female's side between mating and egg laying in order to prevent other males from mating with his bride. The male uses pincers on his abdomen to hang on to the female's head and the pair flies together looking for a good place to lay their eggs.

The common blue damselfly behaves in exactly the same way—so far, nothing out of the ordinary. But as soon as a pair of these damselflies lands on an aquatic plant, something unusual happens. The female marches down the stem, following it under the water like an insect intent on suicide. And it looks as though she is dragging her partner down with her into a watery grave. The male lets go at the last moment, but she marches on without missing a beat. She crawls ever deeper down the plant stem. The brave little insect dives down for up to an hour and a half, reaching depths of up to six feet (two meters) below the surface.

What Dumont was witnessing was not, of course, self-inflicted insecticide, and he was also not the first biologist to observe the remarkable behavior of the female common blue damselfly. But he was the first to film it. Dumont had to return thirty or forty times until he had captured every aspect of this stage of the damselfly's life cycle.

The female damselfly turns into a deep diver so she can deposit her eggs in plant stems underwater. What happens next unfolds according to a predictable pattern. She crawls down a bit, drills a hole, deposits an egg, crawls down a bit farther, drills another hole, and deposits another egg, and so on. Sometimes the little insect seamlessly switches from one plant to another using her abdomen like a tail fin to propel herself through the water.

The female damselfly doesn't suffocate during her long journey to lay her eggs, even though, unlike her larvae, she doesn't have gills. Instead, she encloses herself in a thin bubble of air that performs in a similar way. The oxygen she needs is transferred from this trapped layer of air into the damselfly's breathing holes. Although the oxygen in her small air pocket is quickly depleted, she doesn't need to return to the surface. Once the concentration of oxygen in her bubble is less than that of the water that surrounds her, oxygen molecules stream from the water into her bubble. Physicists call this process diffusion; biologists also refer to the damselfly's trick as "physical gill respiration."

But why does she go to all this effort? We can only speculate. Perhaps this is how common blue damselflies avoid being attacked by larger dragonflies—escaping potential conflicts over coveted spots to lay eggs at the water's surface. It could also be that her courageous dives ensure that her valuable eggs don't dry out if water levels in her wetland habitat fall during dry summers.

When her last egg has been laid, the female damselfly simply lets go. The air in her breathing organs, the trachea, allows her to float like an angel right back up to the surface. But her story is not over yet. Water tension at the surface traps her, making it impossible for her to simply fly away. She flaps helplessly at the spot where she resurfaces.

Her struggles alert dangerous water striders, predatory insects in the order Hemiptera, to her presence. They speed across the surface of the water on their thin legs and attack the helpless female. Luckily, the male damselflies make sure her story has a happy ending. They patrol the surface like little helicopters. They are constantly on the lookout for females they can mate with, and they drive off the hungry water striders with angry aerial attacks.

Attracted by the flapping of the female's wings, a male lands on his chosen one, grabs her with the pincers on his abdomen, and begins to beat his wings with all the power he can muster. Finally, both insects achieve liftoff and the female is saved.

A thrilling story with a happy ending that you can observe in a river meander or gravel pit near you.

~~~~~~

ONE MORNING I received a message from a large supermarket chain: "Save the bees!" In an effort to show it was doing the right thing, the discount chain wanted to let me know it was partnering with the University of Hohenheim and the Heinz Sielmann Foundation to plant acres of wildflower meadows around more than thirty of their distribution centers to help save bees and other insects. I have no idea what the overall ecological impact of the chain might be, but what does interest me is that in 2019 insects were suddenly poster children for a supermarket advertising campaign. Perhaps that is a good sign—and it means that how people are viewing insects has radically changed. In the words of Senegalese forestry engineer Baba Dioum:

> In the end we will conserve only what we love, we will love only what we understand, and we will understand only what we are taught.

I began my book thinking I might change people's minds about insects. Perhaps many of the closed doors I was hoping to open have already been opened wide. I very much hope so.

Acknowledgments

Without my mentor, Professor Martin Dambach, I would never have developed my love for insects. After many discussions with and much advice from him, I turned my hand to writing. I also owe him thanks for his support and advice as this book took shape.

I thank the communities of Yunguilla, San Pablo de Kantesiya, San Francisco de las Pampas, and the *Fundación Otonga*. They welcomed me with open arms to work in their forests, as a scientist and as a journalist.

I thank Professor Giovanni Onore for his unstinting efforts to preserve the last natural paradise in Ecuador, and also for his hospitality and support on all my journeys through Ecuador.

Santiago Molina and Marisol Ayala were the first to tell me of Yumbo's story and the comeback of the spectacled bear in the area around Quito. Many thanks.

I thank Professor Serge Dumont for everything I learned as a result of his work on gravel pits, great crested grebes, and damselflies.

Many thanks to Anke Klüter for all the ideas about jellyfish and cnidarians.

Thomas Weidenbach, many thanks for years of journalistic collaboration and countless discussions about why we write and why we make films in the first place.

I thank my parents, Gitta and Wolfgang, not only for enduring my desire to investigate the natural world but also for actively encouraging it.

My constructive critic Petra Sperling carefully reviewed the manuscript as it took shape, and for that, dearest Petra, I am very, very grateful.

Endnotes

1   Karl von Frisch, *Twelve Little Housemates*, enlarged and revised edition, trans. A. T. Sugar (Oxford: Pergamon, 2015), 76.

2   Eugene Garfield, "The Cockroach Connection—Ancient, Seemingly Indestructible Pest," *Current Comments* 33, no. 45 (1990).

3   Garfield, "The Cockroach Connection."

4   T. Holstein and P. Tardent, "An Ultrahigh-Speed Analysis of Exocytosis: Nematocyst Discharge," *Science* 223, no. 4638 (1984): 830–833, https://doi.org/10.1126/science.6695186.

5   Sanchari Banerjee et al., "Structure of a Heterogeneous, Glycosylated, Lipid-Bound, *In Vivo*-Grown Protein Crystal at Atomic Resolution From the Viviparous Cockroach *Diploptera punctata*," *IUCrJ* 3, no. 4 (July 2016): 282–293, https://doi.org/10.1107/S2052252516008903.

6   J. H. Fabre, *The Hunting Wasps*, trans. Alexander Teixeira de Mattos (New York: Dodd, Mead, 1919), 124.

7   Fabre, *The Hunting Wasps*, 124–125.

8   Fabre, *The Hunting Wasps*, 81–83.

9   Fabre, *The Hunting Wasps*, 84, 86.

10   As early as 1976, in his book *The Selfish Gene*, the British biologist Richard Dawkins focused on genes

as the central place where evolution works. Ultimately, every life-form has only one goal: to preserve and pass on its genes. It is not concerned with the preservation of the species itself.

11 Regen also went by the name Johann, which is why this article is under that name. J. Regen, "Über die Anlockung des Weibchens von Gryllus campestris L. durch telephonisch über-tragene Stridulationslaute des Männchens," *Pflüger's Arch.* 155 (1913): 193–200, https://doi.org/10.1007/BF01680887.

12 Mónica I. Retamosa Izaguirre, Oscar Ramírez-Alán, and Jorge De la O Castro, "Acoustic Indices Applied to Biodiversity Monitoring in a Costa Rica Dry Tropical Forest," *Journal of Ecoacoustics* 2, no. 1 (2018): tnw2np, https://doi.org/10.22261/jea.tnw2np.

13 Terry L. Erwin, "Tropical Forests: Their Richness in Coleoptera and Other Arthropod Species," *The Coleopterists Bulletin* 36, no. 1 (1982): 74–75.

14 Latin America's Indigenous peoples refer to themselves as *indígenas*. They reject the expression "Indians." "Indigenous peoples" is, of course, a general term. There are many differ-ent groups with different cultures and different languages across Latin America. In Amazonia, the locals proudly speak of *naciones*, that is to say, individual nations such as the Cofán, the Shuar, the Huaorani, and many more.

15 Associated Press, "What Americans Heard in Cuba Attacks: The Sound," YouTube video, 1:58, October 12, 2017, https://www.youtube.com/watch?v=rgbnZG85IRo.

16 Alexander L. Stubbs and Fernando Montealegre-Z, "Recording of 'Sonic Attacks' on U.S. Diplomats in Cuba Spectrally Matches the Echoing Call of a Caribbean Cricket," *bioRxiv* (2019), https://doi.org/10.1101/510834.

17 Zhiyuan Shen, Thomas
R. Neil, Daniel Robert, et
al., "Biomechanics of a
Moth Scale at Ultrasonic
Frequencies," PNAS 115,
no. 48 (2018):
12200–12205, https://
doi.org/10.1073
/pnas.1810025115.

18 Harmonic overtones are
tones whose frequency is
an integral multiple of a
fundamental tone. The field
cricket *Gryllus campestris*,
for example, sings with a
carrier frequency of around
5,000 hertz (oscillations
per second). Its harmonic
overtones, therefore, have
frequencies of 10,000,
15,000, 20,000 hertz.

19 F. Nischk and D. Otte, "Bio-
acoustics, Ecology, and
Systematics of Ecuador-
ian Rainforest Crickets
(Orthoptera: Gryllidae:
Phalangopsinae), With
a Description of Four
New Genera and Ten
New Species," *Journal of
Orthoptera Research*, no. 9
(2000): 229, https://doi
.org/10.2307/3503651.

20 Justin O. Schmidt,
The Sting of the Wild
(Baltimore: Johns Hopkins
University Press, 2016),
225.

21 Eric R. Eaton, "Wasp
Wednesday: *Pepsis grossa*,"
Bug Eric (blog), November
21, 2012, http://bugeric
.blogspot.com/2012/11
/wasp-wednesday-pepsis
-grossa.html.

22 There are myriad relationships between bromeliads
and insects. This U.S. study
gives a fascinating over-
view: J. H. Frank and L. P.
Lounibos, "Insects and
Allies Associated With
Bromeliads: A Review,"
*Terrestrial Arthropod
Reviews* 1, no. 2 (2009):
125–153, https://doi
.org/10.1163/18749830
8X414742.

23 C. W. Rettenmeyer, M. E.
Rettenmeyer, J. Joseph, et
al., "The Largest Animal
Association Centered on
One Species: The Army
Ant *Eciton burchellii*
and Its More Than 300
Associates," *Insectes Soci-
aux* 58 (2011): 281–292,
https://doi.org/10.1007
/s00040-010-0128-8.

24 Rettenmeyer et al., "The Largest Animal Association Centered on One Species."

25 Harry C. Evans, Simon L. Elliot, and David P. Hughes, "Hidden Diversity Behind the Zombie-Ant Fungus *Ophiocordyceps unilateralis*: Four New Species Described From Carpenter Ants in Minas Gerais, Brazil," *PLoS One* 6, no. 3 (2011): e17024, https://doi.org/10.1371/journal.pone.0017024.

26 The University of Florida Book of Insect Records (http://entnemdept.ufl.edu/walker/ufbir/), edited by Thomas J. Walker, is a collection of numerous top performances from the world of insects. Which insects are the largest and heaviest or the smallest and lightest? Which have the most offspring and which live the longest?

27 The report admits that the number is only an estimate. The number of endangered insect species is the most uncertain.

28 Rachel Carson, *Silent Spring*, 40th anniversary edition (Boston: Houghton Mifflin, 2002), 2.

29 Carson, *Silent Spring*, 108.

30 Richard Inger, Richard Gregory, James P. Duffy, et al., "Common European Birds Are Declining Rapidly While Less Abundant Species' Numbers Are Rising," *Ecology Letters* 18, no. 1 (2014): 28–36, https://doi.org/10.1111/ele.12387.

31 Diana E. Bowler, Henning Heldbjerg, Anthony D. Fox, et al., "Long-Term Declines of European Insectivorous Bird Populations and Potential Causes," *Conservation Biology* 33, no. 5 (2019): 1120–1130, https://doi.org/10.1111/cobi.13307.

32 M. Sorg, H. Schwan, W. Stenmans, and A. Müller, "Ermittlungen der Biomassen flugaktiver Insekten im Naturschutzgebiet Orbroicher Bruch mit Malaise-Fallen in den Jahren 1989 und 2013," *Mitteilungen aus dem Entomologischen Verein Krefeld* 1 (2013): 1–5.

33 Francisco Sánchez-Bayo and Kris A. G. Wyckhuys, "Worldwide Decline of the

Entomofauna: A Review of
Its Drivers," *Biological Conservation* 232 (2019): 8–27,
https://doi.org/10.1016
/j.biocon.2019.01.020.

34 Bradford C. Lister
and Andres Garcia,
"Climate-Driven Declines
in Arthropod Abundance
Restructure a Rain-
forest Food Web,"
PNAS 115, no. 44 (2018):
E10397–E10406,
https://doi.org/10.1073
/pnas.1722477115.

35 Kristofer M. Helgen, C.
Miguel Pinto, Roland Kays,
et al., "Taxonomic Revision
of the Olingos (*Bassari-
cyon*), With Description of a
New Species, the Olinguito,"
ZooKeys 324 (2013): 1–83,
https://doi.org/10.3897
/zookeys.324.5827.

List of Species